D0978678

TECHNOLOGY-SUPPORTED
MATHEMATICS LEARNING ENVIRONMENTS

SIXTY-SEVENTH YEARBOOK

WILLIAM J. MASALSKI

Sixty-seventh Yearbook Editor
University of Massachusetts
Amherst, Massachusetts

PORTIA C. ELLIOTT

General Yearbook Editor
University of Massachusetts
Amherst, Massachusetts

NATIONAL COUNCIL OF
TEACHERS OF MATHEMATICS

Library of Congress Cataloging-in-Publication Data

Technology-supported mathematics learning environments / William J. Masalski, sixty-seventh yearbook editor ; Portia C. Elliott, general yearbook editor.
   p. cm. -- (Sixty-seventh yearbook)
 Includes bibliographical references.
 ISBN 0-87353-569-3
1. Mathematics--Study and teaching--Technological innovations. I. Masalski, William J. II. Elliott, Portia C. III. National Council of Teachers of Mathematics. IV. Yearbook (National Council of Teachers of Mathematics) ; 67th.
 QA1.N3 67th
 [QA11.2]
 510 s--dc22
 [510/.

2005000006

The National Council of Teachers of Mathematics is a public voice of mathematics education, providing vision, leadership, and professional development to support teachers in ensuring mathematics learning of the highest quality for all students.

The publications of the National Council of Teachers of Mathematics present a variety of viewpoints. The views expressed or implied in this publication, unless otherwise noted, should not be interpreted as official positions of the Council.

Printed in the United States of America

Contents

    REBECCA MCGRAW
        University of Arizona
        Tucson, Arizona
    MAUREEN GRANT
        North Central High School
        Indianapolis, Indiana

    GEORGE C. REESE
        University of Illinois at Urbana-Champaign
        Champaign, Illinois
    JENNIFER DICK
        Daniel Wright Junior High School
        Lincolnshire, Illinois
    JAMES P. DILDINE
        University of Illinois at Urbana-Champaign
        Champaign, Illinois
    KATHLEEN SMITH
        Champaign Central High School
        Champaign, Illinois
    MIKKEL STORAASLI
        West Leyden High School
        Northlake, Illinois
    KENNETH J. TRAVERS
        University of Illinois at Urbana-Champaign (Emeritus)
        Champaign, Illinois
    SUSAN WOTAL
        Daniel Wright Junior High School
        Lincolnshire, Illinois
    DALIA ZYGAS
        West Leyden High School
        Northlake, Illinois

    BOB COULTER
        Missouri Botanical Garden
        Saint Louis, Missouri
    JOSEPH J. KERSKI
        U.S. Geological Survey
        Denver, Colorado

PART 3: QUESTIONS ABOUT THE FUTURE

M. KATHLEEN HEID
The Pennsylvania State University
University Park, Pennsylvania

Preface

Technology is playing an increasingly important role in the teaching and learning of mathematics at all levels. As indicated in the NCTM *Principles and Standards for School Mathematics* and as stated in the NCTM position paper entitled "The Use of Technology in the Learning and Teaching of Mathematics," "Technology is an essential tool for teaching and learning mathematics effectively; it extends the mathematics that can be taught and enhances students' learning." Page 1 of this yearbook contains the full text of the NCTM position paper. The term *technology*, as used in this yearbook, refers to all forms of electronic devices, including computers, calculators and other handheld devices, telecommunications equipment, and the multitude of multimedia hardware, including the software applications associated with their use. These devices and related applications continue to be less and less distinguishable in both form and function, as does one's ability to categorize them and their appropriate use in the teaching and learning process.

To foster the establishment or improvement of technology-supported mathematics learning environments, Part 1: How Research Informs (articles 1–8) reports on overviews of research and findings on the impact of technology on the learning and teaching of mathematics, K–16. These research findings should help guide instruction and assessment decisions made by teachers.

As the twenty-first century unfolds, technology is becoming more visible in mathematics classrooms. Teachers are feeling its impact as they try to capture the vision of NCTM's Technology Principle in their classrooms. But finding effective ways to use technology for teaching, learning, and assessing mathematics can still be a daunting task. The articles in Part 2: Notes from the Field (articles 9–22) are intended to furnish a rich context in which to observe teachers in prekindergarten through grade 12 and teacher educators using technology to help their students better understand mathematics. From its opening paper, which addresses the reality of using technology in the classroom, to its concluding article, which showcases how GIS technology can be used to transform the mathematical landscape, this section offers insight for mathematics educators to use both established and emerging technologies to effectively enhance the teaching and learning of mathematics.

The concluding section of this yearbook, Part 3: Questions about the Future (article 23), gives us all a glimpse of what the future might hold in store for us. In "Technology in Mathematics Education: Tapping into Visions of the Future," the author seeks answers to the following questions: What will technology look like in mathematics classrooms ten years from now? How will technologies be used to affect school mathematics content and teaching? What are the principal changes in mathematics teaching and learning that we might expect over the next decade? The goal of this final article is to look forward to what technology-supported mathematics learning environments might look like in the future. To find answers to this question, the author interviewed twenty-two individuals, all of whom are noted for leading-edge thinking on the uses of technology, most of them in the use of technology in the teaching and learning of mathematics.

This yearbook consists of two separate but closely related pieces: the printed yearbook and an accompanying CD. The CD includes electronic features that enhance an understanding of the articles presented in the printed yearbook. URLs indicated in the hardcopy can be found in click-on Web-accessible format on the CD, as can files illustrating the software used in the articles themselves. Web access to trial versions of the software considered in the printed yearbook is also available directly from the CD. Worksheets and other related material are included, as appropriate. Articles 2, 9, 15, 16, 17, 21 and 23 can also be found in their entirety on the CD, since they each have many Web references.

This yearbook could not have come into existence were it not for the great effort and commitment to this project from the authors of the individual papers, from the Editorial Panel, from the stellar staff at the NCTM Reston office, and—most important—from Portia C. Elliott, general editor for the NCTM yearbooks published in 2005 through 2007. In addition to the general editor and the Sixty-seventh Yearbook editor, the Sixty-seventh Yearbook Editorial Panel consisted of the following individuals:

Suzanne Alejandre	The Math Forum
Glen Blume	The Pennsylvania State University
Ihor Charischak	Stevens Institute of Technology
Mary Ann Connors	Westfield State College
Enrique Galindo	Indiana University

William J. Masalski
Sixty-seventh Yearbook Editor

Note: Numerous Web links have been referenced in this volume. These links were active and correct at the time of the production of this publication. NCTM cannot verify that all these links will remain active over time.

PROLOGUE

NCTM POSITION PAPER ON TECHNOLOGY

An NCTM position statement on …

The Use of Technology in the Learning and Teaching of Mathematics

Position

Technology is an essential tool for teaching and learning mathematics effectively; it extends the mathematics that can be taught and enhances students' learning.

Rationale

Calculators, computer software tools, and other technologies assist in the collection, recording, organization, and analysis of data. They also enhance computational power and provide convenient, accurate, and dynamic drawing, graphing, and computational tools. With such devices, students can extend the range and quality of their mathematical investigations and encounter mathematical ideas in more realistic settings.

In the context of a well-articulated mathematics program, technology increases both the scope of the mathematical content and the range of the problem situations that are within students' reach. Powerful tools for computation, construction, and visual representation offer students access to mathematical content and contexts that would otherwise be too complex for them to explore. Using the tools of technology to work in interesting problem contexts can facilitate students' achievement of a variety of higher-order learning outcomes, such as reflection, reasoning, problem posing, problem solving, and decision making.

1

Technologies are essential tools within a balanced mathematics program. Teachers must be prepared to serve as knowledgeable decision makers in determining when and how their students can use these tools most effectively.

Recommendations

- Every school mathematics program should provide students and teachers with access to tools of instructional technology, including appropriate calculators, computers with mathematical software, Internet connectivity, handheld data-collection devices, and sensing probes.

- Preservice and in-service teachers of mathematics at all levels should be provided with appropriate professional development in the use of instructional technology, the development of mathematics lessons that take advantage of technology-rich environments, and the integration of technology into day-to-day instruction.

- Curricula and courses of study at all levels should incorporate appropriate instructional technology in objectives, lessons, and assessment of learning outcomes.

- Programs of preservice teacher preparation and in-service professional development should strive to instill dispositions of openness to experimentation with ever-evolving technological tools and their pervasive impact on mathematics education.

- Teachers should make informed decisions about the appropriate implementation of technologies in a coherent instructional program.

(October 2003)

1

Teaching Strategies for Developing Judicious Technology Use

Lynda Ball

Kaye Stacey

THIS paper is written as a response to the frequent calls to ban the use of calculators and computer software that "do mathematics" (especially arithmetic or algebra) in schools or at particular grade levels. Our position is that banning the use of technology is not necessary, since with careful teaching, students can learn to make sensible choices about when to use and when not to use technology, and consequently their mathematical power is enhanced. Other authors (e.g., Berry 2002; Heid and Edwards 2001; NCTM 2000; Fey et al. 1995) have persuasively presented the reasons why increased mathematical power can arise from technology use (reasons that include increased opportunities for learning, increased opportunity for real-world problem solving, and orientation to the future), and it is not our intention to repeat these arguments or the evidence for them here. Instead, we shall describe the teaching strategies that teachers can use to produce students who are judicious users of technology and illustrate these strategies by showing the approach of one expert teacher. These strategies need to be well known, so that they auto-

---

Editor's note: The CD accompanying this yearbook contains a hyperlinked URL for a Web site that is relevant to this paper.

matically guide teaching with mathematical technology, whether in the form of a calculator or a computer.

Although the general strategies we describe are applicable to scientific and four-function calculators, this paper is written specifically for classrooms using computer algebra systems (CAS). After many years of struggle defining an appropriate response to the availability of calculators for doing arithmetic, teachers and curriculum designers now need to respond to the widespread availability of CAS, which have the capacity to work with algebraic symbols as well as with numbers and graphs. Recently, Heid (2002) summarized the arguments currently presented by opponents of the use of CAS in American schools. There are practical arguments about cost, ease of use, and inadequate preparation for unchanged college courses and tests. More fundamentally, however, there is the expectation that allowing students to use symbolic manipulation of algebraic expressions, functions, and equations will lead to an unacceptable loss of pen-and-paper algebra skills and that students will reach for the technology to carry out the most elementary of procedures. With good teaching, CAS use need not have this result (Heid 1988) but instead can create judicious users of technology with increased mathematical power. The purpose of this paper is to describe what this teaching is like. There is no need to avoid the use of technology if teachers attend to this issue.

Before we begin, we pause to describe the mathematical behavior of a student who uses technology well for mathematical work. Our ideal user is able to move easily between mathematical notation and technology syntax and can use the tools available flexibly and effectively. Our ideal student has a strong conceptual understanding of mathematics and has sufficiently strong basic algebraic manipulation skills to solve simple problems quickly either mentally or by using paper-and-pencil techniques. However, we see large variations in the expected paper-and-pencil algebraic manipulation skills of our ideal student: in some settings students need extensive, highly developed skills and in others not. This is a local decision.

It is important that our ideal student is one who uses technology *judiciously*. He or she routinely considers whether it is or is not efficient to use technology to solve a problem, not reaching for the technology when a little thought can quickly give an answer. An ideal student also appreciates how technology can be used in an exploratory way, to suggest hypotheses and to test them systematically.

The opposite of judicious use is the "fishing" or "zapping" behavior shown by some students who press buttons seemingly at random with just a vague hope that something might work. A tendency for some students to work in this way is an international problem that good teaching can address

(Pierce and Stacey 2004; Guin and Trouche 1999). Research by Pierce (2002) indicates that some students are always judicious users and others persist with passive or random, unthinking use. However, she found that a large, middle group can be helped to learn to work judiciously, even over a one-semester course.

A Classroom Vignette

This section illustrates how one teacher focused on developing judicious use of technology with a group of fourth-year high school students who were taking a college preparatory course in Australia. The vignette has been constructed from an amalgam of typical classroom observations of Lucy,[1] one of four teachers who were taking part in a research study to examine how the curriculum, teaching, and assessment might change when some forms of CAS are available to students (see Computer Algebra Systems in Schools Curriculum, Assessment & Teaching Project at extranet.edfac.unimelb.edu.au/ DSME/CAS-CAT). Lucy is recognized by colleagues within and beyond her own school as an expert user of many technologies for doing mathematics. In this project, her class had access to CAS calculators for all work during the final two years of high school, including all assessment. This was the first opportunity Lucy had had to teach with CAS always available. Lucy's expertise with this particular technology enabled her to respond to technical queries from her students easily, and we saw in our research project that this ability allowed her to focus on integrating technology use with the development of mental strategies and paper-and-pencil techniques. She often stated that her main goal for the use of technology was improving the mathematical understanding of students, not the learning of technology.

The vignette takes place during a unit on exponential and logarithmic functions, when Lucy poses the problem in figure 1.1.

The vignette demonstrates the teaching strategies that Lucy implemented (see fig. 1.2) to help her students become judicious users of technology. For Lucy, this was a major goal that guided many of her actions:

> I do notice that some weaker students are drawn to using it [technology] before they think [about what method may be most efficient],... [and] the teaching issue is "How do you make them think before they do it?" You keep revisiting it, making comments such as, "Okay, you could do that [use tech-

1. Lucy is not her real name. The teacher's anonymity has been protected by changing the name and possibly the gender. Quotes from Lucy have been edited to shorten them and increase readability. The authors wish to thank Lucy and her students for their generous participation in the project and for teaching us so much.

Fish in a lake are dying of an unusual disease. The population was 10000 one week after the first dead fish was discovered, and 8000 after 5 weeks. Assume that the number of fish can be modelled by the equation $p(t) = a \ln(t) + b$, where t is the number of weeks after the first dead fish was found.	1. Find the values of a and b. 2. Find the predicted number of fish after 10 weeks. 3. Find when the population is predicted to be 3000. 4. Sketch a graph of $p(t)$ against t.

Fig. 1.1. The problem presented to Lucy's class

nology], but it would take you five minutes to do it. Here's a way [mental strategies or pencil and paper] that would take you thirty seconds but where you just have to think a bit first." … Brighter students are more like the teacher. They are more discerning and more versatile with the choice of techniques they might use.

The importance that Lucy gave to mathematical understanding meant that she taught technical skills in the context of mathematical problems (see Step 1 in fig. 1.2). As an overall approach, she followed what is called a "white box–black box" approach (Buchberger 1990), teaching new procedures first by pencil and paper and then later with technology. Altogether, she adopted a reasonably conservative approach in expecting students to develop a solid foundation of pencil and paper manipulative skills. However, Lucy has a curriculum where technology is completely integrated, and so she would often demonstrate both approaches, with a discussion of the benefits of each. In the vignette in figure 1.2, Lucy demonstrated a technology technique (Step 5) before its pencil and paper parallel (Step 6).

My personal preference is that for every new procedure students see it first by hand [pencil and paper]. They learn to do it by hand for simple examples, and then progressively, as the manipulations get more difficult by hand, or when manipulation is not the core part of the problem, you … encourage the students to use the calculator to do it. I also impress on the students that in questions where the manipulation is just part of a bigger question, they should do it in the quickest or most efficient way possible. If that's the calculator, then use the calculator, and if it's not the calculator, then don't use it.

Lucy intended that students should make a habit of carefully inspecting a problem to get an overview, think about how the solution might proceed, and consider the nature of the answer. In Steps 1, 2, and 3 in figure 1.2, Lucy placed a tactical restriction on technology use, requesting that students draw on their understanding of the features of families of curves to sketch a graph. By insisting that the graph be sketched manually, Lucy sent an implicit message that students should spend some time at the beginning of a problem, not using procedures but instead drawing on their algebraic insight to prepare. Students are then in a position to make conscious decisions about when to reach or not reach for technology and to know what to expect from the technology output. Lucy intended that this become a habit for her students.

Of course, the class discussion in Steps 1, 2, and 3 was also a primary way of helping the students develop the algebraic insight that is needed to work with technology (Pierce and Stacey 2002). For example, through her consideration of the different graphs produced by students Lucy was able to discuss important features, resulting in the class's being able to identify the features that were crucial to either accepting or rejecting each graph. This skill is extremely important in a classroom with access to technology where students will need to interpret technology outputs in the context of given problems. Lucy's strategy to restrict technology use in Steps 1, 2, and 3 is thus intended to be beneficial for students when they are using technology. By having some expectation of technology outputs, they may be able to identify errors, for example, those arising in the entry syntax.

Lucy integrated the use of technology, mental strategies, and paper-and-pencil techniques in her teaching, thereby legitimizing them all. However, she often made definitive statements about when students should or should not use technology. She encouraged students to think about the extent and nature of technology use, so that they would become judicious users of technology, making informed decisions about appropriate methods for solving problems. Throughout Steps 5, 6, and 7 and later in the individual discussions in Step 8, Lucy emphasized when technology use is an efficient and effective method for solving problems and when it is unnecessary. In fact, Lucy commented that this emphasis was perhaps the greatest change in her teaching that had resulted from having constant access to CAS technology over two years.

> I'm making more overt statements that students should choose the most efficient method.… I would have done that [in the past with other technology], but using CAS introduces a really strong [invitation to] "Do it this way." But I have found that this is not necessarily the best method. [Now] I think about whether it's the best tool and which ways within the CAS are the best ways to do it.… That's probably the big thing that's changed [in my teaching].

In the vignette from Lucy's class, there is a complex interplay among technology, mental strategies, and pencil and paper techniques. In this short

Step 1

Lucy presented the problem in Figure 1.1 and asked students to sketch a graph of the situation. Lucy asked students not to use technology but instead to consider the likely shape of the graph based on their experience with logarithmic functions and the given information.

Sample Student Graph

Sample Student Graph

Step 2

Lucy had a number of students draw their graphs on the board, commenting that she would not say whether their graphs were correct or incorrect at this stage.

Step 3

Lucy led a class discussion about graphs of logarithmic functions before examining the graphs sketched by students. The class discussed why each graph might or might not give an appropriate representation of the information given in the problem.

Step 4

Students were asked to find the values of *a* and *b* as quickly as possible. Students could decide whether to use pencil and paper or technology. Technology users needed to select appropriate features (e.g., substitute or solve simultaneously) to find the two constants.

Fig. 1.2. Vignette demonstrating how Lucy integrates the use of mental, pencil-and-paper, and technology approaches

Step 5

While students were finding the function, Lucy entered her solution into the calculator, switching on the overhead projector display when students had finished. Lucy stored the original function and used the built-in "solve" feature to find the two values. Storing the function using function notation made the subsequent equation solving easy for Lucy. Lucy's focus was on solving the problem, not teaching calculator syntax.

> Lucy models effective and efficient use of technology.

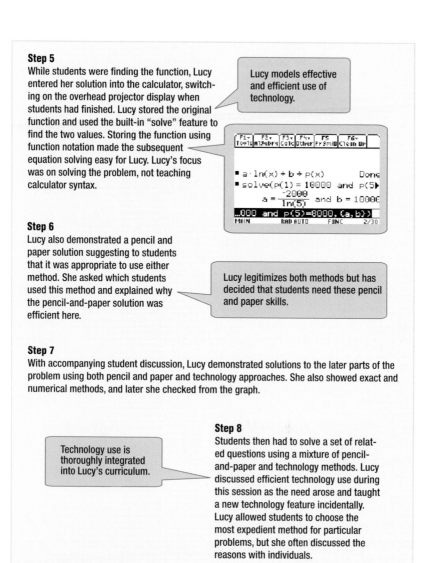

Step 6

Lucy also demonstrated a pencil and paper solution suggesting to students that it was appropriate to use either method. She asked which students used this method and explained why the pencil-and-paper solution was efficient here.

> Lucy legitimizes both methods but has decided that students need these pencil and paper skills.

Step 7

With accompanying student discussion, Lucy demonstrated solutions to the later parts of the problem using both pencil and paper and technology approaches. She also showed exact and numerical methods, and later she checked from the graph.

> Technology use is thoroughly integrated into Lucy's curriculum.

Step 8

Students then had to solve a set of related questions using a mixture of pencil-and-paper and technology methods. Lucy discussed efficient technology use during this session as the need arose and taught a new technology feature incidentally. Lucy allowed students to choose the most expedient method for particular problems, but she often discussed the reasons with individuals.

> Curriculum and assessment are aligned.

Step 9

Later, on the class test, Lucy included some questions that were best done by paper and pencil and some that were not. Equal credit was awarded for solutions done with both pencil and paper and technology.

Fig. 1.2 (continued)

sequence Lucy discourages technology use (Step 1) with the intention of developing students' understanding, discussing why this is occurring as she does this. However, she also demonstrates technology use and teaches new technology skills in the context of the same problem (Steps 5 and 8). She allows students to choose whether or not they use technology to solve a problem, focusing on efficiency and speed (Step 4) and encouraging students to use CAS judiciously. Regularly making choices and discussing in class the efficiency of different approaches will most likely increase students' metacognitive awareness. In Lucy's situation, the class discussion is usually led by the teacher, orchestrating comments from the students about their current work.

Teaching Strategies for Helping Students Become Judicious Technology Users

The classroom vignette in figure 1.2, which is typical of Lucy's approach, contains examples of four general teaching strategies to help students become discerning users of technology. These four strategies are summarized in figure 1.3, along with some suggestions on how these general strategies might be implemented in different classrooms and at different times. Lucy is an excellent example of a teacher who pays attention to judicious use of technology, but hers is not the only way. Other teachers in our project, for example, have used these strategies rather differently.

Promote Careful Decision Making about Technology Use

The first general strategy is to promote students' metacognition, their ability to think about their thinking. In this context, metacognition is concerned with students' knowledge and beliefs about appropriate roles for mental strategies, pencil and paper techniques, and technology use in doing mathematics and is also concerned with the way in which students regulate and control their mathematical work (Weinert 1987). As in Lucy's classroom, decisions about whether to solve a problem using mental strategies, pencil and paper, or technology—or a combination—need to become a topic of class discussion. The teacher can share his or her reasons for deciding to use or not use technology with the class, and students can discuss their decisions with one another and compare. Often these decisions will be based on the complexity of individual operations, but good decision making will also often require looking ahead in a problem to anticipate the solution path through a series of steps of varying complexity.

A feature of good problem solvers' metacognition is that they can usually assess the difficulty of a problem accurately in relation to their own skills (Krutetskii [1962] 1969). This ability is essential for judicious technology use.

Teaching Strategy	Classroom Suggestions
Promote careful decision making about technology use	• The teacher shares decision making about mental, pencil-and-paper, and technology approaches with class. • Students discuss decisions about mental, pencil-and-paper, and technology approaches with one another. • Students monitor their own underuse or both overuse of technology. • Teacher and class discuss the nature of human and technological contributions.
Integrate technology into the curriculum	• Include examples that really require technology use. • Include examples where technology use would be inefficient. • Generally expect students to make their own choices of mental, pencil and paper, or technology approaches. • Focus on mathematics rather than on syntax.
Tactically restrict the use of technology for a limited time	• Ban technology use for a few minutes or a day. • Ban a technology feature for a few minutes or a day. • Use temporary bans to develop pencil-and-paper skills, an appreciation of the role of technology, algebraic insight, or a specific concept.
Promote habits of using algebraic insight for overview and monitoring	• The teacher models this routinely at the start of each problem. • The teacher encourages the continual monitoring of technology outputs against expectations.

Fig. 1.3. Four teaching strategies for developing judicious technology users

Students need to appreciate that it is usually more efficient to solve simple problems mentally or by using pencil and paper techniques, and this is a clear benefit from developing strong basic pencil and paper skills and good algebraic insight. However, those with fewer skills can minimize the likelihood of errors by compensating with technology. An instructive exercise to promote careful decision making is to have students monitor their own overuse or underuse of technology on an assignment. Research on assessment using graphing calculators by Boers and Jones (1994) and by Forster and Mueller (2002) generally shows that students underuse technology even when it is freely available.

Much of the heat of debate about permitting technology in schools arises from the public's perception of the nature of mathematics. When students discuss the relative contributions of themselves and technology to solving a problem, they will quickly identify how the technology can neither formulate a problem mathematically nor interpret the results. Experience with technology also highlights the limitations and idiosyncrasies of even the best systems, since they are all only computer programs that carry out certain operations in specified circumstances. Students who have the opportunity to discuss these issues can develop a broader appreciation of what it means to do mathematics.

Integrate Technology into the Curriculum

An environment where students are usually given the responsibility to make their own selections of mental, pencil and paper, and technology approaches furnishes the best opportunity for training for judicious use of technology. Teachers need to be able to guide students to become good users, not teach as if technology did not exist and hope that students will adjust well by themselves when the tools are permitted. In order to do this, technology needs to be integrated into the curriculum as Lucy does, not ignored or added as an afterthought.

A curriculum where technology use is well integrated is characterized by a mix of problems, some that are best done without technology and others that really demonstrate its power. If the curriculum includes only questions that are comfortably within the range of expected paper-and-pencil skills, then permitting technology sends the message that technology is either useless or (worse) should be used on simple problems. Similarly, assessment will encourage the discriminating use of technology if it contains some items that are most efficiently done with technology and some without it.

Tactically Restrict the Use of Technology for a Limited Time

Imposing a temporary ban on technology use is a useful option available to teachers, even if they generally permit students free access to technology and integrate it fully into the curriculum. In order for a temporary ban to be effective, both the teacher and the students need to be clear about its purpose. Sometimes teachers will ban technology use for a set of problems so that students will practice their mental and pencil and paper skills. Lucy, the teacher in the classroom vignette in figure 1.2, regularly bans technology use for a few minutes to develop her students' algebraic insight and the habit of thinking ahead to the type of results they will expect before they rush in to use either pencil and paper or technology techniques. We do not advocate banning technology use for a whole topic, since this discourages the curriculum integration that is essential for developing judicious use.

Inventive teachers can also use other selective bans to good effect, creating activities analogous to the "broken calculator" activities that teachers have been using for many years with four-function calculators. For example, a teacher may tell students to imagine that the "4" key on their calculator is broken, yet they are still to do a variety of calculations, such as $4086 - 499$, on the calculator. Students might calculate $5086 - 500 - 1000 + 1$; or they might even calculate $5186 - 599$, adding 1100 to the first term and 100 to the second term and then subtracting 1000 from the difference. In this way, the activity challenges students to rename numbers and to use properties of subtraction. One analogous activity for algebra beginners using CAS is to solve equations without using the "solve" commands. In this way CAS can be used to produce a sequence of equivalent equations, thereby strengthening students' understanding of "do the same to both sides" without their being derailed by algebraic errors. Another activity of this type is to ban temporarily the use of the built-in "solve" features for simultaneous equations, thus requiring students to devise a strategy to solve using other features, such as substitution. Another suggestion is that students could try to show the principal features of a graph using the fewest "zooms."

Good classroom discussion of the appropriate role of technology can be stimulated by prohibiting half the class from using any technology for a day and requiring the other half to use it for absolutely everything that day. If the problems are chosen well, neither side will be content. Good discussion can follow.

Promote Habits of Using Algebraic Insight

Just as strong number sense is essential for doing arithmetic, whether assisted by pencil and paper or by technology, so strong algebraic insight is essential for doing mathematics, whether with pencil and paper or with CAS technology. Algebraic insight is a mix of conceptual knowledge and simple procedural knowledge that is needed to supplement, guide, and monitor procedural work. Important elements of this insight include identifying the structure and key features of expressions and being able to link representations, such as the easy conversion between symbols and graphs. A detailed elaboration is given by Pierce and Stacey (2002).

The explicit use of algebraic insight should become a habit so that when students see an algebraic expression, they think about what they already *know* about its symbols, structure, and key features and possibly its graph before they move further into the question. Researchers have found (Pierce and Stacey 2004; Pierce 2002) that if teachers demonstrate this initial step routinely in class, it is likely to become a habit for their students. Lucy's first step in the vignette shows how she does this. Algebraic insight is also essential to monitor technology output throughout a solution process, so that students need to be encouraged to constantly ask themselves whether their results are reasonable. In her individual interactions with students, this is a question that Lucy frequently asks.

Conclusion

We have selected an example of teaching from Lucy's class because she felt very strongly the need to develop judicious technology use and considered that her own growth in teaching throughout our three-year project was mainly in this regard. Since she began as an expert CAS user (although without prior experience of consistently using CAS in class), she was able to direct her attention to this issue and to try out ideas, gradually creating a familiar classroom routine. Other teachers in our project also approached this issue thoughtfully, and although they used similar general strategies, the details of implementation were distinctly different. One of them almost never used temporary tactical restrictions. As in any other aspect of teaching, there are many different ways of being effective.

The purpose of this paper has been to offer four general strategies (outlined in fig. 1.3) that teachers can use to help their students become judicious users of technology. There is much more to say about learning to teach with technology, but the issue of having students become judicious users is of high importance to all teachers, whether technology is a small or large part of their mathematics program. The desired attention to paper-and-pencil techniques may be quite variable from one setting to another for the foreseeable future, but all students with any access (current or future) to technology need to become judicious users. Furthermore, we believe that teachers who attend to this issue can confidently deliver programs that result in students using technology to increase their mathematical power instead of randomly pressing buttons in the hope that an answer may appear by magic.

REFERENCES

Berry, John. "Developing Mathematical Modeling Skills: The Role of CAS." *Zentralblatt für Didaktik der Mathematik* 34 (October 2002): 212–20.

Boers, Monique A., and Peter L. Jones. "Students' Use of Graphics Calculators under Examination Conditions." *International Journal of Mathematical Education in Science and Technology* 25 (July-August 1994): 491–516.

Buchberger, Bruno. "Should Students Learn Integration Rules." *SIGSAM Bulletin 24* 91, no. 1 (1990): 10–17.

Fey, James T., M. Kathleen Heid, Richard A. Good, Glendon W. Blume, Charlene Sheets, and Rose Mary Zbiek. *Concepts in Algebra: A Technological Approach.* 2nd ed. Chicago: Everyday Learning, 1995.

Forster, Patricia, and Ute Mueller. "Assessment in Calculus in the Presence of Graphics Calculators." *Mathematics Education Research Journal* 14 (November 2002): 16–36.

Guin, Dominique, and Luc Trouche. "The Complex Process of Converting Tools into Mathematical Instruments: The Case of Calculators." *International Journal of Computers for Mathematical Learning* 3, no. 3 (1999): 195–227.

Heid, M. Kathleen. "Resequencing Skills and Concepts in Applied Calculus Using the Computer as a Tool." *Journal for Research in Mathematics Education* 19 (January 1988): 3–25.

———. "Computer Algebra Systems in Secondary Mathematics Classes: The Time to Act Is Now!" *Mathematics Teacher* 95 (December 2002): 662–67.

Heid, M. Kathleen, and M. Todd Edwards. "Computer Algebra Systems: Revolution or Retrofit for Today's Classrooms?" *Theory into Practice* 40 (spring 2001): 128–36.

Krutetskii, Vadim A. "An Analysis of the Individual Structure of Mathematical Abilities in School Children." In *The Structure of Mathematical Abilities,* vol. 2, Soviet Studies in the Psychology of Learning and Teaching Mathematics, edited by Jeremy Kilpatrick and Izaak Wirszup, pp. 59–104. 1962. Reprint, Chicago: University of Chicago, 1969.

National Council of Teachers of Mathematics (NCTM). *Principles and Standards for School Mathematics.* Reston, Va.: NCTM, 2000. Available at standards.nctm.org/document/chapter2/index.htm.

Pierce, Robyn. "An Exploration of Algebraic Insight and Effective Use of Computer Algebra Systems." Ph.D. diss., University of Melbourne, 2002.

Pierce, Robyn, and Kaye Stacey. "Algebraic Insight: The Algebra Needed to Use Computer Algebra Systems." *Mathematics Teacher* 95 (November 2002): 622–27.

———. "A Framework for Monitoring Progress and Planning Teaching towards Effective Use of Computer Algebra Systems." *International Journal of Computers for Mathematical Learning* 9, no. 1 (2004): 59–94.

Wienert, Franz E. "Metacognition and Motivation as Determinants of Effective Learning and Understanding." In *Metacognition, Motivation and Understanding,* edited by Franz E. Weinert and Rainer H. Kluwe, pp. 1–19. Hillsdale, N.J.: Lawrence Erlbaum, 1987.

2

Young Children's Use of Virtual Manipulatives and Other Forms of Mathematical Representations

Patricia S. Moyer

Deborah Niezgoda

John Stanley

For those who teach mathematics, using technology in the mathematics classroom has meant the increased use of Geometer's Sketchpad, spreadsheets, and graphing calculators. Although these technology applications are most often used in middle school and high school mathematics, many technology tools are appropriate for younger learners. An exciting technology for use in the early grades is called *virtual manipulatives*. Three elements have converged to form this class of manipulative technology. Those elements include innovations in computer technology that allow programmers to create electronic objects, the availability of Internet resources, and educators' recognition of the usefulness of concrete manipulatives and other representations for teaching mathematics. As a result of these advances, computer programmers created *virtual manipulatives*.

A virtual manipulative is defined as "an interactive, Web-based visual representation of a dynamic object that presents opportunities for constructing mathematical knowledge" (Moyer, Bolyard, and Spikell 2002, p. 373). Virtual manipulatives are often exact visual replicas of concrete manipulatives (such as pattern blocks, base-ten blocks, geometric solids, Cuisenaire rods, or geoboards). They are placed on the Internet as applets—smaller, stand-alone versions of application programs. Children can use the computer mouse to manipulate the images.

Teachers and researchers are discovering the useful properties of virtual manipulatives (Moyer and Bolyard 2002) and measuring their impact on stu-

The authors would like to thank the members of the project team, Linda Garrett, Spencer Jamieson, Lori Knotts, Lindsay Sweetser, and Allison Ward, Westlawn Elementary School, Falls Church, Virginia.

Editor's note: The CD accompanying this yearbook contains an electronic version of this article, complete with hyperlinked URLs.

dents' learning (Reimer and Moyer 2003). This paper discusses virtual manipulatives as a unique technological form of representation for teaching mathematics and describes how two elementary school teachers conducted action research projects in their classrooms using this technology. During these projects, a kindergarten teacher and a second-grade teacher taught a series of lessons and documented children's use of two virtual manipulatives (virtual pattern blocks and virtual base-ten blocks). In the process, the teachers gathered data on the children's use of technology and examined how it enhanced the children's understanding of mathematical concepts and skills.

Virtual Manipulatives: A Unique Technological Representation

In the National Council of Teachers of Mathematics (NCTM) *Principles and Standards for School Mathematics* (NCTM 2000), virtual manipulatives fall under the Process Standard *Representation.* Although Representation is new as a process strand to the NCTM *Principles and Standards,* educators have used different forms of representation in teaching mathematics for many years. The most common forms of representation used in school mathematics include concrete/physical representations (manipulatives and 3D geometric models), pictorial/visual representations (pictures, drawings, and other visual images), and abstract/symbolic representations (numbers, letters, and operation signs).

In addition to these familiar representations used in mathematics, forms of representation have begun to appear on the Web. Some of these representations are static and are essentially pictures that can be viewed but not manipulated. (For example, a picture of pattern blocks with questions printed below the blocks would be a static representation.) These might be called "virtual math activities" because they appear on the Web, but they would *not* fit the definition of virtual manipulatives.

Other representations on the Web are dynamic, visual objects. Many of them look like concrete manipulatives and can be moved as one would move an object. The ability to move, or manipulate, these visual objects makes them interactive for children and makes them "virtual manipulatives." Children who interact with this different form of technology will have *different* mathematical experiences. As the Technology Principle indicates, "Work with virtual manipulatives ... can allow young children to extend physical experience and to develop an initial understanding of sophisticated ideas like the use of algorithms" (NCTM 2000, pp. 26–27).

Because virtual manipulatives are a unique mathematics tool, one might wonder if virtual manipulatives are an additional form of representation. If we consider them in relation to the three forms of representation currently used in school mathematics (physical, pictorial, symbolic), virtual manipulatives do not fit neatly into any one of these three categories. Virtual manipulatives do furnish a visual image like a pictorial model, and yet they can be moved and manipulated like a physical model. In essence, they have some of the advantageous properties of both of these forms of representation, as well as some additional advantages brought about by their technological properties. Because of these attributes, virtual, concrete, and pictorial representations support one another in the mathematics teaching and learning process. Each representational form contributes important information that supports a more complete understanding of underlying mathematical ideas. Therefore, one should not be a substitute for another.

Virtual manipulatives are uniquely suited for teaching mathematics with young children. A Web connection makes them free of charge and easily *available.* Some virtual manipulatives have the *potential for alteration.* For example, users can color parts of objects; they can mark the sides of a polygon or highlight the faces on a Platonic solid. This *interactivity* allows all children to be engaged in the problem-solving process. Some virtual manipulatives *link symbolic and iconic notations* by saving numerical information or providing mathematics notations that label the on-screen objects. The click of the mouse on many virtual manipulatives gives children access to *unlimited materials.* And *clean up is easy* — children simply click an icon and the on-screen objects disappear.

Numerous virtual manipulative Web sites are currently available. Some of the largest collections of virtual manipulatives can be found at the NCTM Web site (www.nctm.org/), the National Library of Virtual Manipulatives (www.matti.usu.edu/nlvm), and the Shodor Education Foundation (www.shodor.org/).

Teaching Mathematics with Virtual Manipulatives

The classroom projects in this paper were conducted at Westlawn Elementary School in Falls Church, Virginia—a Title I school that serves a linguistically and economically diverse population. Just thirty minutes from the nation's capital, the school demographics include a population of students 75 percent nonwhite with 34 percent second-language learners. In each of the classes in the following vignettes, the children had some prior experiences with technology and with the virtual manipulative Web sites during mathematics lessons. Therefore, they were able to use the virtual manipulatives efficiently for the mathematical tasks the teachers designed.

Two of the authors of this article are classroom teachers, and they led instruction during each of the class sessions. Five observers recorded anecdotal notes on the children's work during the mathematics interactions. Each observer was assigned to a small number of children (three or four) so that the observer could record the children's work and ask questions throughout the lessons. Children's work was analyzed and condensed for this article to highlight major points in each of the lessons.

Kindergarten Patterns

The full-day kindergarten class was composed of eighteen children, twelve of whom spoke a language other than English in their homes. The group was also ethnically diverse, including Hispanic, Asian, African-American, Middle Eastern, and Caucasian students. During the three-day lesson on patterns, the children focused on both repeating and growing patterns of many kinds. For example, the children had previous experiences with simple repeating patterns (i.e., ABB, AAB, or ABAB) and had been introduced to growing patterns (i.e., A, AB, ABC, ABCD, or 1, 1-2, 1-2-3, 1-2-3-4). Children had explored patterns visually, aurally, and kinesthetically using shapes, colors, sounds, and rhythms. They also investigated patterns in literature, music, and nature. The ability to recognize, describe, and extend patterns is an essential mathematics Standard for grades K–2 (NCTM 2000).

On the first day of their three-day lesson on patterns, the children used wooden pattern blocks to make their patterns and a long strip of oaktag paper as a work mat. On the second day, the children used virtual pattern blocks to create their patterns. Virtual pattern blocks are available on several Web sites, including NCTM (nctm.org), Arcytech (www.arcytech.org/), and the National Library of Virtual Manipulatives (www.matti.usu.edu/nlvm). (Fig. 2.1 shows a child's pattern on the Arcytech Web site.) On the third day, the children created and drew patterns freehand on 4″ × 18″ strips of construction paper using crayons or markers.

To analyze the children's work, the project team examined how the form of representation (wooden pattern blocks, virtual pattern blocks, or drawings) influenced the variety and complexity of the children's patterns. The results that follow are summarized in table 2.1.

The team first compared the number of individual patterns each child constructed on each day. This comparison showed that the children made a greater number of patterns using the virtual pattern blocks than they did drawing the patterns or using the wooden pattern blocks. (See table 2.1.) An analysis of the variety of pattern types the children made each day indicated that the most common patterns were AB, ABB, AAB, ABC, and ABCD across all

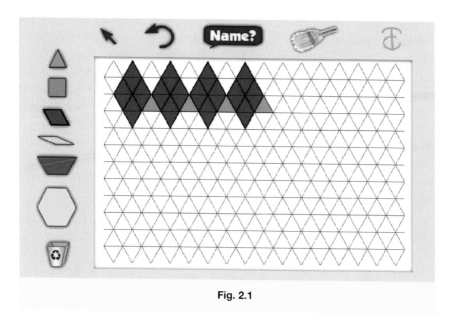

Fig. 2.1

three forms of representation. This analysis also indicated that the pattern types children made each day remained consistent. For example, those who created ABC patterns on Day 1 also made ABC patterns on Days 2 and 3.

Table 2.1
Quantitative Summary of Pattern Results — Kindergarten

Pattern Characteristics	Day 1 Wooden Pattern Blocks	Day 2 Virtual Pattern Blocks	Day 3 Children's Drawings
Total # of Patterns Created	48 pat/18 st Avg. = 2.7 pat/st	63 pat/18 st Avg. = 3.5 pat/st	46 pat/17 st Avg. = 2.7 pat/st
# of Blocks in Each Pattern Stem	51.33 bl/18 st Avg. = 2.9 bl/stm	59.75 bl/18 st Avg. = 3.3 bl/stm	48.17 bl/17 st Avg. = 2.8 bl/stm
# of Blocks Used During Patterning	783 bl/18 st Avg. = 43.5 bl/st	1019 bl/18 st Avg. = 56.6 bl/st	609 bl/17 st Avg. = 35.8 bl/st
Avg. # of Blocks Used to Make Each Pattern	783 bl/48 pat Avg. = 16.3 bl/pat	1019 bl/63 pat Avg. = 16.2 bl/pat	609 bl/46 pat Avg. = 13.3 bl/pat

Note. N = 18; one student was absent on Day 3.
Note. pat = patterns; st = student(s); bl = blocks/elements

To judge the complexity of the children's patterns, the team examined the number of elements the children used in their pattern stems. For example, an AB pattern has two elements, an AABB pattern has four elements, and an ABCD pattern also has four elements. (Fig. 2.2 shows a pattern with four elements in the pattern stem.) We found that three children used only two or three elements in their pattern stems every day, whereas all other children used four, five, six, eight, or even ten elements in their pattern stems. (Fig. 2.3 shows a pattern with ten elements in the pattern stem.) This analysis also showed that children used more elements in their pattern stems using the virtual pattern blocks than they did when drawing patterns or using the wooden pattern blocks. (See table 2.1.)

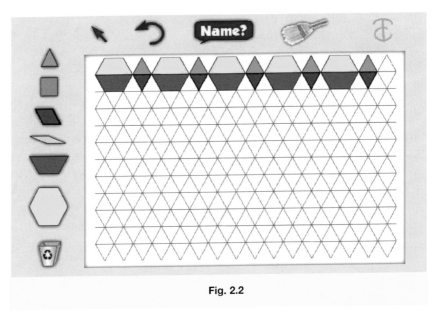

Fig. 2.2

Another examination looked at the total number of blocks the children used in each pattern. This analysis highlighted the work of children who may have completed a small number of patterns but who repeated those patterns numerous times. (Fig. 2.4 shows a child's pattern with numerous repeats.) This number was consistent with the number of patterns the children created. For example, there were more blocks incorporated on the day the children used the virtual pattern blocks than on the days they drew patterns or used wooden pattern blocks. However, when we examined the number of patterns created in relationship to the number of elements used to create those patterns, the number of elements in each pattern was consistent between virtual pattern blocks and wooden pattern blocks and slightly less for the drawings of blocks. (See table 2.1.) Children

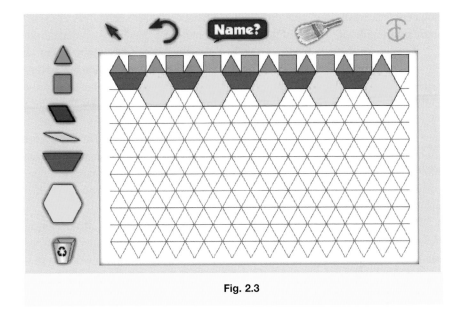

Fig. 2.3

seemed to lose interest in the pattern they were creating when they reached a certain number of blocks (approximately thirteen to sixteen) or repeats.

Another analysis examined patterns that went beyond simple repeating, left-to-right, horizontal patterns. The team classified several patterning

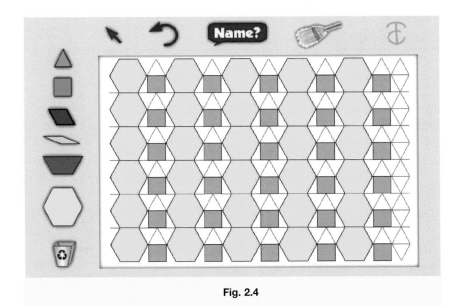

Fig. 2.4

behaviors as exhibiting "creativity." These creative pattern behaviors are summarized in table 2.2. Some of these creative pattern behaviors included standing the wooden pattern blocks on end or stacking them to create tessellating patterns. (Being able to stack and build with the blocks is one of the advantages of using the wooden blocks in making patterns.) More than half of the children used the virtual pattern blocks to create new shapes by using one block to partially cover another. (Fig. 2.5 shows how a child used a hexagon to cover a rhombus and create a blue triangle.) Other creative behaviors included making vertical patterns or tessellations with the virtual pattern blocks. (Fig. 2.6 shows a child's tessellation.) Children also made growing patterns with each different form of representation. (Fig. 2.7 shows a child's growing pattern with the virtual pattern blocks.)

The final analysis examined overall similarities and differences among the three representational forms. We found that children created approximately the same number of patterns when drawing or using the wooden pattern blocks, but they created a greater number of patterns with the virtual pattern blocks. They used more elements in the pattern stem and exhibited more behaviors that the team classified as creative when using the virtual pattern blocks. Overall, the team identified several advantages for the children when they had the opportunity to create the patterns using the virtual pattern blocks. However, each form of representation allowed these kindergarten learners many opportunities to communicate the complexity of their thinking.

Table 2.2
Creative Pattern Behaviors—Kindergarten

Pattern Behaviors Classified as "Creative"	Day 1 Wooden Pattern Blocks	Day 2 Virtual Pattern Blocks	Day 3 Children's Drawings
Standing Blocks on End	6 st		
Creating a New Shape		11 st	
Vertical Pattern		2 st	
Tessellating Pattern	6 st	9 st	
Growing Pattern	2 st	2 st	4 st
Totals	14	24	4

Note. $N = 18$; one student was absent on Day 3.
Note. st = student(s)

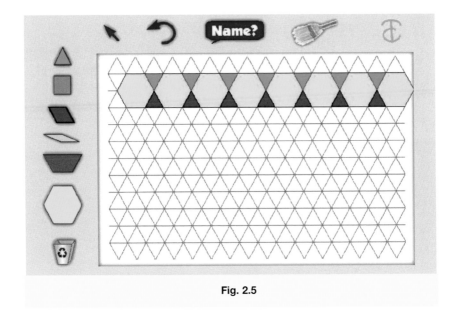

Fig. 2.5

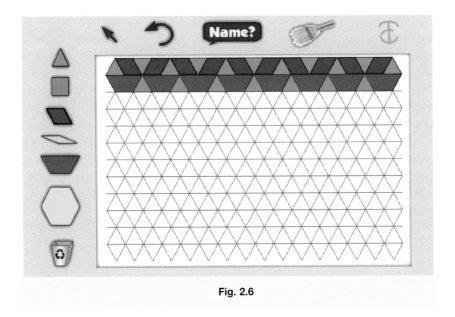

Fig. 2.6

Second-Graders Regrouping with Virtual Base-Ten Blocks

The second-grade class was composed of nineteen children, fifteen of whom spoke a language other than English in their homes. The group includ-

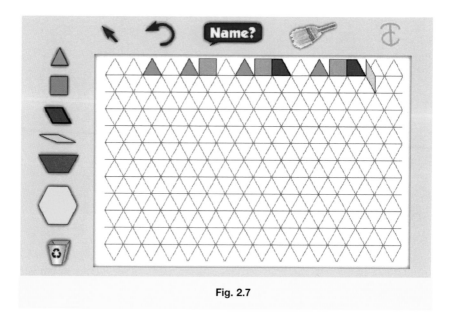

Fig. 2.7

ed Hispanic, Asian, and Caucasian students. During the two-day lesson on addition, the children focused on the regrouping of one- and two-digit numbers using virtual base-ten blocks, drawings, and written explanations. The class had previous experiences using the wooden base-ten blocks for number operations, and the use of these blocks was very effective for learning the process of trading ones blocks for tens blocks to model regrouping. The teacher was particularly interested in finding out if manipulating the virtual base-ten blocks on Day 1 would enhance the children's abilities to develop their visual-representation skills and to make drawings of the regrouping process on Day 2. The ability to understand ways of representing numbers and meanings of operations is an important part of the Standards for second-grade mathematics (NCTM 2000).

On Day 1 of the lesson, children used virtual base-ten blocks. Virtual base-ten blocks are available on several Web sites, including NCTM (nctm.org), Arcytech (www.arcytech.org/), and the National Library of Virtual Manipulatives (www.matti.usu.edu/nlvm). The virtual base-ten blocks are an electronic replica of the concrete base-ten blocks with representations of hundreds, tens, and ones blocks on a place-value mat. Children can group ten ones to make a ten by gluing them together, and they can break apart a tens rod to make ten ones. As the children used the virtual base-ten blocks, they recorded the sums for several one- and two-digit addition exercises. (Fig. 2.8 shows a child's work representing 68 + 15.) On

Day 1, many of the children lined up the ones cubes in a straight line, counted them to confirm there were ten, glued ten ones to make a tens rod, and moved the newly created tens rod to the tens column. This process was observable and allowed all children, particularly those who were second-language learners, to demonstrate and explain the process they were using to the observers.

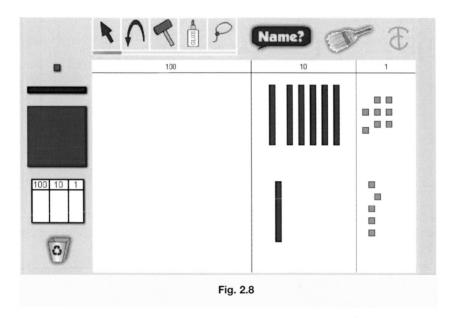

Fig. 2.8

On Day 2, the children received four pieces of paper, each with a different addition exercise. Each paper had one addition exercise, space for the children to draw a solution, and room to provide a written explanation. The teacher encouraged children to draw a representation of the addition processes on their papers for each addition exercise. The team was particularly interested in how the children would draw and explain their processes for regrouping ones to form a ten. To solve the problem 46 + 6, one child wrote: "I put 4 in the 10s place and 6 in the ones place and I put 6 more in the ones place. It was 12. I put 10 in a circle. I put the ten in the tens place." (Fig. 2.9 shows her drawing.)

As observers monitored the children's drawings during class, they noted that children circled ten ones on their papers and used arrows to simulate moving ten ones to the tens place as they had done the previous day with the virtual base-ten blocks. (Fig. 2.10 shows a child's drawing with arrows.) Some of the children also drew a new tens rod on their papers to show that ten ones became

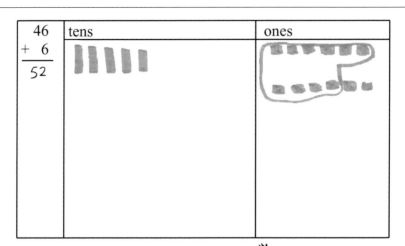

This is how I solved the problem: I put 4 in the 10's place and put to 6 in the one's place and I put 6 more in the one's place. It was 12. I put 12 in a circle. I put the ten in the tens place.

Fig. 2.9

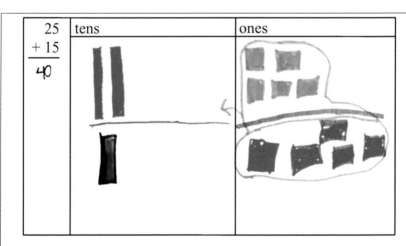

This is how I solved the problem:

I Solved This I had 3 Tens and 10 ones.

Fig. 2.10

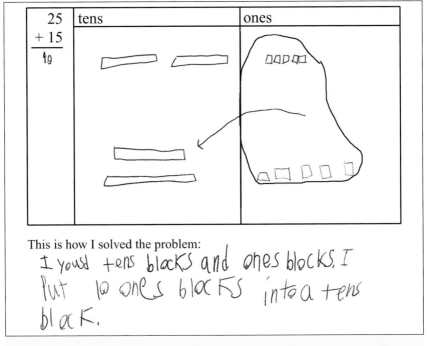

Fig. 2.11

one ten. To explain this, one child wrote, "I used tens blocks and ones blocks. I put 10 ones blocks into a tens block." (Fig. 2.11 shows his drawing of a new tens rod.) When solving the addition exercise 25 + 15, one child wrote, "I drew 3 tens then I put two fives at ones. I knew it was 40 because there were 30 then 5 + 5 = 10. So you turn the fives into a ten so it's 40. Goodbye." The children's drawings and explanations seemed to show that the virtual base-ten models offered a visual image the children could use to demonstrate the addition process on Day 2. Their explanations also indicated that they were developing flexibility with composing and decomposing numbers.

To analyze the children's work, the project team examined how the children explained and represented the concept and procedures for addition with regrouping. We calculated the percents of the children's correct answers on both days. These results are summarized in table 2.3. On Day 1, children worked in the computer lab and completed a worksheet with twelve addition exercises. The children completed 83 percent of the twelve addition exercises on the worksheet with 75 percent accuracy. On Day 2, children worked in the classroom and completed four addition exercises. For each exercise the children wrote the symbolic answer, drew a picture showing the addition

Table 2.3

Addition Problem Completion Rates and Strategy Use – Grade 2

	Day 1 Virtual Base-10 Blocks	Day 2 Drawings & Explanations
Problem Completion		
Number of Exercises Given	12	4
Percent of Exercises Completed	83%	80%
Accuracy of Symbolic Computation	75%	75%
Accuracy in Drawings/Explanations	NA	88%
Strategies Used by Students		
Counting	2	0
Counting & Place Value	6	0
Place Value	7	13
Place Value & Algorithm	4	6

Note. N = 19

process, and wrote an explanation for their drawings and symbols. The children completed 80 percent of the four addition exercises with 75 percent accuracy. However, they completed the drawings and written explanations with 88 percent accuracy. The reason for the discrepancy between the accuracy of the exercises only and the accuracy of the drawings and explanations was because many of the children drew an accurate picture of the addition exercise and wrote an accurate explanation for the problem; however, they did not complete the entire task by writing the sum for the addition exercise. Others drew an accurate representation with a correct explanation and wrote the wrong sum.

The team was also interested in whether the children used a counting method, a place-value-based regrouping method, or some other strategy. To document this process, the observers talked with children and examined their writings and drawings on both days. This part of the analysis indicated that on Day 1 (using virtual base-ten blocks), two children were simply counting; six children used both counting and a place-value strategy; seven children used only a place-value strategy; and four children used a place-value strategy with the traditional algorithm. On Day 2, thirteen children used a place-value strategy, and six children used a place-value strategy with the traditional algorithm. (See table 2.3.) For example, the child's work and written explanation in figure 2.12 shows that she understands the difference between numbers in the tens and ones places. Her explanation states, "I [took] 2 ten(s) to make 20, then I put 3 on the one(s). Then I look. I put 8 on the one(s). Then

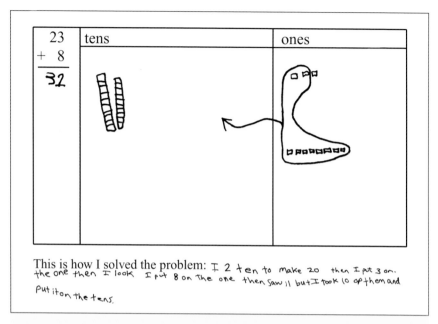

This is how I solved the problem: I 2 ten to make 20 then I put 3 on. the one then I look I put 8 on the one then Saw 11 but I took 10 of them and Put it on the tens.

Fig. 2.12

I saw 11, but I took 10 of them and put it on the tens." We believe that the visual model (the virtual base-ten blocks) helped the children to "see" the regrouping of the numbers during the addition process, which gave this process more meaning for the children.

The team noticed unique methods children used to solve the exercises. On Day 1 (virtual base-ten blocks), three of the children used eleven ones for the numeral 11 instead of using a single tens rod and a single ones unit. Another unique method used by three of the higher-ability children was a more efficient grouping strategy. Using the Web site's grouping tool, called a lasso, the children generated groups of ones, lassoed the ones, and then moved them to the ones column as a group instead of moving one block at a time as most of the other children had done. (Fig. 2.13 shows a child using the lasso tool on the Web site.) Some children used the tens rod as a reference by lining up ones units next to it to confirm that they had ten ones. (Fig. 2.14 shows how a child lines up ones against the tens rod.)

On Day 2 (drawings and explanations), children represented the regrouping process in unique ways. For example, some children used two different colors to represent ones and tens. One child even used two colors for two different sets of ones to show how the ones joined to make one ten. (Fig. 2.15

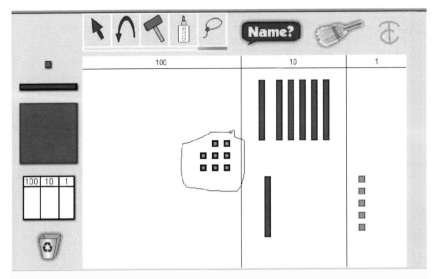

Fig. 2.13

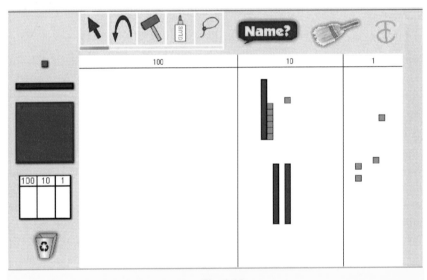

Fig. 2.14

shows this representation for 39 + 23.) The tens and ones for the number 39 are drawn in pink and the tens and ones for the number 23 are drawn in blue. To show the 9 ones and the 1 one joining together to make a ten, the child drew the newly created ten in both pink and blue. We believe that these unique

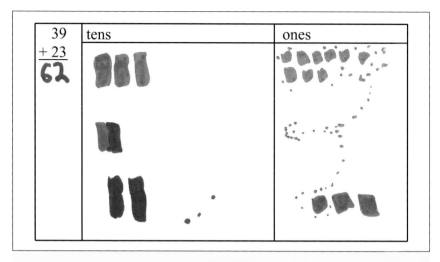

Fig. 2.15

drawings and explanations showed the extent of children's understanding of the addition process.

Research on children's understanding of place value indicates that many children think that the 1 in 15 means one until third or fourth grade (Kamii 1980; Ross 1986). During this project, the visual images of tens and ones seemed to help the children develop meaning for numbers in the ones and tens places. During the weeks following this two-day lesson, the second-grade teacher noticed the relative ease with which the children transferred their understanding to nonpictorial addition exercises.

Closing Comments

The teachers in these projects used the virtual manipulatives to make important connections among (1) the concrete manipulations of objects, (2) the visual images of those objects, (3) abstract mathematical ideas and notations, and (4) the processes underlying these concepts. The virtual manipulatives provided the children with flexible representations that allowed them to explore their ideas. The virtual environment enabled the children to test their mathematical ideas as they were working with the virtual tools. Young children, particularly second-language learners, may have difficulty verbalizing their mathematical thinking. In these classroom projects, using virtual manipulatives helped children clarify their own thinking through experimentation and allowed them to communicate and demonstrate their mathematical thinking to others.

In these classrooms, teachers used the concrete manipulatives (pattern blocks and base-ten blocks) prior to using the virtual manipulatives. The use of the virtual blocks during these projects provided an important bridge for the children between the use of concrete objects and the use of pictorial and symbolic notations for representing abstract mathematical ideas. Some virtual manipulatives offer concept tutorials that link objects and manipulations on those objects with the abstract symbols that represent them. Making this link among representational forms explicit is an important feature of using virtual manipulatives with children. Although these projects examined only two classrooms of children, they document important preliminary findings about the usefulness of virtual manipulatives for teaching mathematics with young children. We believe it is important to investigate problems and questions using different technologies and forms of representations in real classrooms and to explore effective ways to use these technologies in teaching mathematics to all children.

REFERENCES

Kamii, Mieko. "Place Value: Children's Efforts to Find a Correspondence between Digits and Number of Objects." Paper presented at the tenth annual symposium of the Jean Piaget Society, Philadelphia, May 1980.

Moyer, Patricia S., and Johnna J. Bolyard. "Exploring Representation in the Middle Grades: Investigations in Geometry with Virtual Manipulatives." *Australian Mathematics Teacher* 58 (March 2002): 19–25.

Moyer, Patricia S., Johnna J. Bolyard, and Mark A. Spikell. "What Are Virtual Manipulatives?" *Teaching Children Mathematics* 8 (February 2002): 372–77.

National Council of Teachers of Mathematics (NCTM). *Principles and Standards for School Mathematics.* Reston, Va.: NCTM, 2000.

Reimer, Kelly, and Patricia S. Moyer. "Third Graders Learn about Fractions Using Virtual Manipulatives: A Classroom Study." Paper presented at the annual meeting of the American Educational Research Association, San Diego, April 2004.

Ross, Susan. "The Development of Children's Place-Value Numeration Concepts in Grades Two through Five." Paper presented at the annual meeting of the American Educational Research Association, San Francisco, April 1986. (ERIC Document Reproduction Service No. ED 273 482)

3

Five Steps to Zero: Students Developing Elementary Number Theory Concepts When Using Calculators

Carolyn Kieran
José Guzmán

Thinking about a topic in mathematics cannot be separated from tasks and techniques in exploring this topic.

—Jean-Baptiste Lagrange, in *Computer Algebra Systems in Secondary School Mathematics Education*

THE research that we have carried out, and which is presented below, describes a calculator-based activity that has been found to encourage the emergence of number-theoretic thinking among students in the middle grades. The study also illustrates how conceptual knowledge can grow, along with the development of techniques, in a technological environment.

The middle-grade student and the development of elementary number-theoretic concepts have received scant attention in much of the technology-based literature thus far. Despite the fact that the National Council of Teachers of Mathematics (NCTM) *Principles and Standards for School Mathematics* argues for students in these grades to be provided with opportunities to "work with whole numbers in their study of number theory" and to engage in tasks involving "factors, multiples, prime numbers, and divisibility" (NCTM 2000, p. 217), little material dealing with these topics is to be found in current textbooks or professional resources. Even less appears on the ways in which com-

We thank the Social Sciences and Humanities Research Council of Canada (Grant #410-99-1515) and CONACYT of Mexico (Grant #I32810-S) for their support of the research described in this paper. We are also extremely grateful to the teachers and students of the two schools that participated in the study.

Editor's note: The CD accompanying this yearbook contains a Word document containing ten activity sheets that are relevant to this article.

puting technology might be harnessed in the development of mathematical thinking related to these concepts.

A Way of Considering the Role of Technology in the Development of Mathematical Thinking

When we consider the role of technology-supported environments in the learning of mathematics, it is important to have a theoretical framework that can be used not only to help design the environment and its tasks but also to interpret students' actions and learning while they are working in the environment. One such framework can be found in the research on the role of symbol-manipulating calculators in the coevolution of conceptual and technical expertise (e.g., Artigue 2002; Guin and Trouche 1999; Lagrange 2000). Central to the notions underlying this research is the Vygotskian-based work of Verillon and Rabardel (1995), who distinguish two distinct ways of viewing tools—as artifacts and as instruments. A tool, which starts out merely as an artifact, becomes an instrument for the user only when he or she has been able to appropriate it for himself or herself and has integrated it fully within his or her activity. However, in this process of transforming the artifact into an instrument, the learner is not just simply learning tool-techniques that permit him or her to respond to given mathematical tasks. Mathematical concepts codevelop while the learner is perfecting his techniques with the tool.

According to Lagrange (2000), who drew upon earlier work of Chevallard (1999), the dialectical interaction occurring in this process involves three components: technique, task, and the theorizing engaged in by the learner. More specifically, Lagrange (pp. 16–17) points out that

> the work of constituting techniques in response to tasks, and of the theoretical elaboration of the problems posed by these techniques, remains fundamental to learning.... The new instruments of mathematical work are of interest ... because they permit students to *develop new techniques that constitute a bridge between tasks and theories.* [our translation and our emphasis]

If techniques constitute a bridge between tasks and the emergence of theoretical knowledge, then it is by looking at the *techniques* that students develop with their technological instruments, in response to certain tasks, that we obtain a window into the evolution of their mathematical thinking. This is a fundamental thesis of our work and of this paper. However, our focus on techniques cannot be separated from the tasks themselves and their role in mathematical learning. We shall say more about tasks and the thinking behind the design of our tasks in the next section.

Having elaborated the triad of task, technique, and theory that serves as our conceptual framework, we now present our study. We begin with a description of the task situation and the environment. This is followed by a report of the different techniques that students developed in response to the task. The final section discusses the mathematical knowledge (i.e., the theory) that emerged in the process of interaction with the technology.

The Task Situation and the Environment

The "Five Steps to Zero" Problem

The task situation was based on the "Five Steps to Zero" problem (Williams and Stephens 1992; see fig. 3.1[1]). Successfully tackling this task situation, with the constraint of using only the whole numbers from 1 to 9, involves developing techniques for converting numbers (prime or composite) into other numbers in the same neighborhood (not more than 9 away from the given number) that have divisors that are as large as possible (but not larger than 9) so as to reach zero in five or fewer steps. In the example illustrated in figure 3.1, the given number of 151—a prime—needs first to be converted into a composite before looking for suitable divisors different from the number itself and 1. As an alternative to the approach displayed in figure 3.1, a student might first subtract 1 to obtain the number 150, namely, $151 - 1 = 150$. This might then be followed by successive divisions involving 2, 3, and 5, that is, $150 / 2 = 75$, $75 / 3 = 25$, and $25 / 5 = 5$, and then by a final subtraction of 5, which yields $5 - 5 = 0$. Since students were encouraged to use as few steps as possible, this problem situation provided fertile ground for learning that, for example, if a number has both a and b as divisors, then it is also divisible by ab.

What Some of the Research Says

Theoretical realizations such as those above are not so obvious to students. Past research has shown that, for example, just because students find, say, that $a \times b = ab$, they cannot then state that ab is a multiple of b without first dividing ab by b (Vergnaud 1988). In another study involving older students, Zazkis and Campbell (1996, p. 542) asked: "Consider the number $M = 3^3 \times 5^2 \times 7$. Is

---

1. Note that all the whole numbers from 1 to 1000, with the exceptions of 851 and 853, can be brought down to zero in five or fewer steps. These two exceptions require six steps—851 being the product of the two primes 23 and 37, and 853 being itself prime. All the numbers in the vicinity of these two, that is, within 9 on either side of 851 and 853, require five steps to reach zero, thus necessitating six steps for these two.

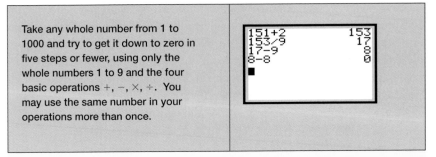

Take any whole number from 1 to 1000 and try to get it down to zero in five steps or fewer, using only the whole numbers 1 to 9 and the four basic operations +, −, ×, ÷. You may use the same number in your operations more than once.

Fig. 3.1. The basic task, "Five Steps to Zero" (based on Williams & Stephens 1992), accompanied by an example (151), displayed on the multiline screen of a graphing calculator

M divisible by 7? Explain. Is *M* divisible by 5, 2, 9, 63, 11, 15? Explain." Students' understanding of the concepts of divisibility and prime decomposition was found to be so poor that the researchers argued that "developing a conceptual understanding of divisibility and factorization [which] is essential in the development of conceptual understanding of the multiplicative structure of numbers" (p. 562) should be happening in the middle grades. We were thus interested in seeing how the tasks and tools that we used with the middle-grades students of our study would increase their awareness of such number-theoretic concepts as factors, multiples, prime numbers, and divisibility.

Tasks Designed around the "Five Steps to Zero" Problem

The study was carried out in seventh-, eighth-, and ninth-grade mathematics classes in Mexico and Canada during two consecutive years. The researchers developed a set of ten activity sheets involving tasks based on the "Five Steps to Zero" problem on which students had worked over a period of one week (five classes of fifty minutes each). (See fig. 3.2 for a summary of the main questions asked; the activity sheets themselves are included on the CD accompanying this volume.)

Each student was equipped with a graphing calculator (some classes used the TI-73 and others the TI-83 Plus); however, the graphing part of the calculator was not used. The larger screen of the graphing calculator, as opposed to the small one-line screen of a simple four-operation calculator, made it possible for students to record and later observe all the steps of their calculations. Even though the students had worked with these calculators before, they had not had extensive experience with them.

In designing the tasks and the sequence in which they were presented, we had a combination of mathematical and metacognitive considerations in

1. Take the number 144. Write as many ways as you can for bringing 144 to zero, using as few steps as possible.

2. Take the number 151. Write as many ways as you can for bringing 151 to zero, using as few steps as possible.

3. Take the number 732. Write as many ways as you can for bringing 732 to zero, using as few steps as possible.

4. Describe your strategies for minimizing the number of steps.

5. Here is a solution proposed by a pupil for bringing 432 to zero: $432/2 = 216$; $216/2 = 108$; $108/2 = 54$; $54/3 = 18$; $18/3 = 6$; $6 - 6 = 0$. Show a way of bringing 432 to zero in fewer steps. Explain your strategy. Do you think it will always work? Why?

6. Here is a strategy proposed by a pupil for bringing 731 to zero: $731 + 7 = 738$; $738/9 = 82$; $82 - 1 = 81$; $81/9 = 9$; $9 - 9 = 0$. Show a way of bringing 731 to zero with fewer steps. Explain your strategy.

7. The number 266 has as its divisors 2, 7, and 19. In other words, $266 = 2 \times 7 \times 19$. What is the best strategy for bringing 266 to zero? Why? Explain why your strategy is best.

8. Here is the strategy proposed by a pupil for bringing 499 to zero: $499 + 1 = 500$; $500/5 = 100$; $100/5 = 20$; $20/5 = 4$; $4 - 4 = 0$. Show a way of bringing 499 to zero in fewer steps. Explain your strategy.

9. What do you consider to be the best strategies for bringing numbers down to zero?

10. Think of a number that your classmates would find difficult to bring to zero in five or fewer steps. Write down why you think it would be a hard number. Show the solution you found for your hard number. (*Note:* This particular task formed the basis for a competition among the three groups of the class, with each group settling on one of the hard numbers proposed by its members; three rounds of competition by two teams trying to solve the hard number of the third team then took place.)

Fig. 3.2. Tasks from the 10 activity sheets prepared to accompany the "Five Steps to Zero" problem

mind. From the mathematical point of view, we began with a simple task—bringing the number 144 to zero in as few steps as possible, 144 being an even number with several divisors below 10. By asking students to record on paper the solutions they tried, we were hoping to have a trace of both their immediate strategies as well as those that evolved as they went along. The next task, which involved the prime number 151, required the addition or subtraction of some number in order to have a composite that could be divided by numbers less than 10. Would students bring 151 down to 144 and apply the strategies already used for the 144 of the previous task, or would they start afresh— and, if so, would they aim for an even number, a number ending in 5 or 0, or something else? The third task began with 732, which could be brought to zero in five steps if one began with a division of 3, 4, or 6; but a four-step solution required adjusting the number so as to have a multiple of 9. After the first three open-ended tasks, followed by the very important fourth task in which students were asked to describe in writing their techniques for minimizing the number of steps to be taken to reach zero, the next task presented a six-step approach for bringing 432 to zero. All of the divisors used in the given example were 2s and 3s. We wanted to see whether students would naturally think to combine some of the given divisors—for example, the first three 2s to yield a divisor of 8, and the last two 3s for a divisor of 9—so as to reduce the number of steps from six to three. In other words, were they developing the realization that dividing once by 2, a second time by 2, and a third time by 2 was the same as dividing by 8? The next task, which was designed to provide some experience with multiples of 9, illustrated a five-step method for bringing 731 to zero that led off with the conversion of 731 to 738. Would students come to see that if 738 is divisible by 9, so too is 738 − 9—and that this would save a step because the resulting quotient of 81 is immediately divisible by 9? The tenth and final task, which was the most open-ended of all for the students, involved thinking of a number that they considered to be difficult to bring down to zero in five or fewer steps. By asking students to write down their "hard" number and to explain why they thought it was a hard one, we were once again attempting to probe the number-theoretic thinking that had evolved for each of them over the course of the week.

The Role of the Calculator in Supporting This Activity

What is it that the calculator makes possible that middle school students would not have access to if they were not using calculators for this activity? The calculator buttons for the operations +, −, ×, and ÷ permit students to carry out these operations in one step. Without having to keep track of all the intermediate moves that would normally capture their attention in a paper-and-pencil environment, they are free to focus on structural aspects that otherwise tend to be eclipsed. By not having to be preoccupied with the details of, for example, the multiplication and division algorithms when they are working in a calculator environment, students can analyze the numerical structure presented by the resulting products and quotients in relation to the numbers that produced them. In this way, the theoretical dimension of the students' work can develop along with the technical strategies, both of which emerge in response to the given tasks. In paper-and-pencil environments, it is much more difficult for students to think about mathematical theory while they are trying to concentrate on obtaining the correct numerical result of an operation.

The Classroom Setup and the Introduction to the Set of Activities

The students of each class were grouped into three teams; however, for most of the activities, they were free to work individually or in smaller groups within their own team. Regularly, individual students were invited to come forward to the classroom's projection screen and to work out a couple of problems on the calculator that was hooked up to it. This allowed both the researchers and the classroom teacher to observe directly the nature of the techniques that students were developing in response to the problem tasks. The week's activities ended with a timed competition among the three teams to see which one could come up with a solution to a "hard number" in the fewest number of steps. During the week that followed the classroom part of the study, four students representing a range of mathematical ability (according to the opinion of the classroom teacher) from each of the participating classes were individually interviewed. This gave the researchers the opportunity to explore at closer range both the mathematical knowledge and the techniques that had evolved over the course of the previous week. All of the classroom activity, as well as the individual interviews that followed, were videotaped.

The classroom mathematics teacher introduced the main task situation as follows. He (or she) began with the example of 360 and illustrated with the projection screen that he (she) could get down to zero as follows: 360/2, 180/2, 90/3, 30/6, 5 − 5. The teacher then requested volunteers to come forward to show how they might get to zero in fewer than five steps. After that, students were invited to suggest their own starting numbers, say, larger than 200, which other pupils came forward to solve. Students then began to work on the tasks on the activity sheets.

The Emergence of Techniques

The techniques that students used at the beginning of the week's activities tended to be based on simple criteria for divisibility, such as dividing by 5 if the number ended in 0 or 5, or dividing by 2 if the number was even. This is illustrated by the work of Marianne, a seventh grader, on the number 151, from the second activity sheet (see figs. 3.3a and 3.3b)[2]. However, the techniques of Marianne evolved, just as they did for her classmates. On the third activity sheet (see figs. 3.4a and 3.4b), she showed a shift toward trying to find the largest divisor possible.

L1 : 151/̶$^{(3)}$ 1		150
L2 : ans/5		30
L3 : ans/5		6
L4 : ans − 6		0

Fig. 3.3a (Marianne)

L1 : $^{(6)}$151/̶$^{(2)}$ +4		155
L2 : ans/5		31
L3 : $^{(2)}$31 − 1		30
L4 : ans */5		6
L5 : ans − 6		0

Fig. 3.3b (Marianne)

L1 : 732/6	122
L2 : 122/2	61
L3 : 61 + 3	64
L4 : 64/8	8
L5 : 8 − 8	0

Fig. 3.4a (Marianne)

L1 : 732/4	183
L2 : 183 − 3	180
L3 : 180/9	20
L4 : 20/5	4
L5 : 4 − 4	0

Fig. 3.4b (Marianne)

On the fourth activity sheet, in describing the techniques that had emerged thus far for her, Marianne wrote:

> Divide by the largest divisor possible from 1 to 9; if there are no divisors, then add or subtract to obtain another number where the division is possible. After dividing, look at the result and test whether division is again possible.

---

2. Legend for calculator-screen transcriptions: Ln—refers to the line of the calculator display screen. Screen lines that have a bar across them denote lines (or characters) that the student has deleted from the screen. The small number in parentheses indicates the time taken by the student before entering the number or operation that follows—5 blinks of the screen cursor being equal to one unit in parentheses.

> If not, repeat the previous procedure until arriving at a number less than 9 and finish the procedure with a subtraction.

Nicolas, an eighth grader, offers us another example of how pupils in this study quickly evolved from more basic techniques to that of trying to find the largest divisor possible. Having tried to see if 931 was divisible by 9 or 8, his next efforts centered on finding a number in the neighborhood of 931 for which he could use large divisors throughout (see figs. 3.5a and 3.5b).

L1 : 931 – 1	930
L2 : 930/9	103.33
L3 : 930/8	116.25
L4 : 930/5	186
L5 : 186/9	20.66
L6 : 186/8	23.25

L26 : 931 + 5	936
L27 : 936/9	104
~~L28 : 10/8~~	~~1.25~~
L29 : 104/8	13
L30 : 13 – 9	4
L31 : 4 – 4	0

Fig. 3.5a (Nicolas)   **Fig. 3.5b (Nicolas)**

Toward the end of the week's activities, several students began to make breakthroughs. Their focus became more controlled in that instead of using successive trial and error with the divisors 9, 8, and 7, they started to search for techniques oriented around the use of the factor 9. Marianne, for example, wrote on her sixth activity sheet that she wanted "to subtract or add in order to arrive at a number divisible by 9; if you divide by the largest number, even if you do a subtraction or yet an addition, you will reach zero more rapidly." With Marianne, it was not until the last day, when she was using the projection-screen device and was given the number 971 to solve, that we witnessed the technique that she had developed (see figs. 3.6a and 3.6b). Marianne began at once to search for a number in the vicinity of 971 by using the product of two factors, one of which was 9. Once she had found two that were on either side of the target number (see lines 8 and 9 of fig. 3.6b), she successively refined her search until she reached a product that was within 9 units of 971 (see line 11).

A technique involving multiples of 9 emerged for another seventh-grade student, Mara, near the end of the week. While she was at the front of the class using the projection screen, a classmate suggested she try 731. After a

L2 : 9×86	774
L3 : [(2)] 9×76	684
L4 : 9×97	873
[What number was it? Someone answered : 971.]	
L5 : $9 \times$ [(3)] 1	
L6 : 9×99	891 [(2)]
[L1, L2, ..., L6]	

Fig. 3.6a (Marianne)

L7 : 9×105	945
L8 : 9×110	990
L9 : 9×107	963
L10 : [(2)] 9×106	954
L11 : 9×108	972
L12 : $971 + 1$	972
L13 : $972/9$	108
L14 : [(3)] $108/6$	18
L15 : $18/9$	2
L16 : $2 - 2$	0

Fig. 3.6b (Marianne)

L1 : 731 [(1)] $+ 1$	732
L2 : $732/9$	81.33
...	
L10 : $731 - 8$	723
L11 : $723/9$	80.33

Fig. 3.7a (Mara)

L16 : $731/9$	81.22
L17 : 9×82	738
L18 : $731 + 7$	738
L19 : $738/9$	82
L20 : $82/2$	41

Fig. 3.7b (Mara)

few unsuccessful tries involving the search for neighboring numbers that could be divided by 9, she seemed suddenly to be struck by the idea of taking the whole-number part of the quotient and doing a reverse multiplication with the next higher integer (see lines 16 and 17 of figs. 3.7a and 3.7b). This told her immediately how much adjustment needed to be done to the initial number. Pablo, an eighth grader, had developed a similar technique (see fig. 3.8).

We return now to Nicolas, whose work with the trial divisors of 9 and 8 on the task involving 931 was shown above (in fig. 3.5). While he was exploring

L1 : 931/9   103.44
L2 : 9 × 103   927
L3 : 931 – 4   927
L4 : ans/9   103 [(5)]
…
L7 : 103/7   14.71

Fig. 3.8 (Pablo)

the task on the third activity sheet, a new thought began to incubate during his fourth try with 732 (see fig. 3.9). On the fourth activity sheet, he described the technique he had used on this task as follows: "First, I multiplied $9 \times 9 \times 9$ and I obtained 729; after that I noticed that $732 - 3$ was 729, and so I did the same operation, but in reverse."

It was during the individual interview with Nicolas, the week following our study, that we were able to see how he had further evolved the technique of multiplying 9s (since no further reference to this technique had appeared subsequent to that which was seen on his fourth activity sheet, described above). When asked what he would do if a given number was not divisible on the first step by a number between 2 and 9, he answered that he would add or subtract. So we continued by asking him how he figured out the amount that he needed to add or subtract,

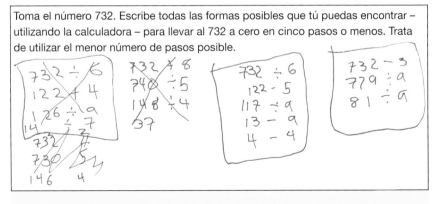

Toma el número 732. Escribe todas las formas posibles que tú puedas encontrar – utilizando la calculadora – para llevar al 732 a cero en cinco pasos o menos. Trata de utilizar el menor número de pasos posible.

Fig. 3.9. Nicholas' activity sheet #3

	Verbatim	Comments
32.	**N:** Because … well, I also have a "technique" that I use. First, I do a multiplication, say, $9 \times 9 \times 3$ or something like that to arrive at another number, and I look at that number.	N. doesn't really answer the question; rather, he gives another way to attack the problem, using the word "technique."
33.	**I:** Let's see, repeat that for me one more time.	Interviewer wants to be sure about what N. has just said.
34.	**N:** For example, if I have the number 571 and I multiply 9×9, it gives 81.	Here N. has chosen his own starting number.
35.	**I:** Let us say that I give you the number 431.	The Int. wishes to offer one for which N. may not have a ready-made solution.
36.	**N:** OK, so I go: L1: 9×9 81 L2: ans $\times 3$ 243 L3: $9 \times 9 \times 4$ 324 ~~L4:~~ ~~9~~ L5: $9 \times 9 \times 5$ 405 So, like that, I arrive more quickly L6: $9 \times 9 \times 5$ 405	We notice here that N. generates three potential factors, two of them being 9s. Then he systematically adjusts one of them. <u>Legend</u> for calculator screen transcription: see Footnote #2.
37.	**I:** But I said 431. With this strategy that you have just described, how do you begin?	Int. attempts to get N. back to the given number 431.
38.	**N:** First, $9 \times 9 \times$ something, no? Until arriving close to the number. For example, L7: $9 \times 8 \times 6$ 432	N. now controls the last two factors at the same time. The 9 is reduced to 8 while the 5 is increased to 6.
39.	**I:** Yes, I told you 431.	Again, Int. wants to bring the task to its conclusion.

Fig. 3.10. Segment of transcript from the interview with Nicolas

to which he responded that he had a certain "technique" (see fig. 3.10 for the transcript of this segment of the interview—I = Interviewer and N = Nicolas).

We note here that Nicolas—in contrast to Marianne, whose newly developed technique involved controlling two factors at a time, one of which was 9— had learned to master three factors. These three factors, which had originally started out as $9 \times 9 \times 9$—as seen on the third activity sheet—had evolved to include several combinations involving other numbers.

Verbatim	Comments
44. **N:** So, 431 plus 1, divided by 6, divided by 8, and so on.	N. states a solution that involves dividing first by the last factor of 9 × 8 × 6, and then moving toward the left, inversing each operation, until arriving at 9, which is handled by a final subtraction.
45. **I:** Let's see.	Int. invites N. to show it on the calculator.
46. **N:** L8: 431 + 1 432 L9: 432/6 72 L10: ans/8 9 L11: ans – 9 0 And there it is!	N. enters his solution into the calculator.

Fig. 3.10 (continued)

Discussion:
Analysis of the Evolution of Mathematical Knowledge

Some of the most powerful mathematical explorations that occurred during the week of activity on the "Five Steps to Zero" task involved the search for multiples of 9. Since students wanted to arrive at zero in the fewest number of steps possible, their initial techniques soon evolved into attempts to discover whether the given number was divisible by 9, or whether any numbers in the close vicinity (i.e., within 9 on either side of a given number) were. But how to find the right numbers in the close vicinity was the question. Since most students were unaware of the criterion for divisibility by 9 (i.e., sum the digits to see if the total is a multiple of 9), they were not able to resort to this technique. In fact, many of those who were aware of the criterion for divisibility by 9 tended to carry it out in a rather mechanical fashion, and so they deprived themselves of the rich conceptual learning that occurred for the majority of the other students.

The mathematical knowledge that emerged for many students involved variants of *the division algorithm*. According to the division algorithm, any whole number can be expressed as the product of two whole numbers plus remainder, that is, "For any $b > 0$ and a, there exist unique integers c and d with $0 \leq d < b$ such that $a = b \times c + d$" (e.g., $989 = 9 \times 109 + 8$). Even though students were not explicitly taught this algorithm or its formalization, their work, which showed the different means by which they tried to obtain the value of c, illustrated the ways in which they were beginning to think about this algorithm. Their struggles to find multiples of 9 in the vicinity of the given numbers also disclosed their emergent discoveries of other related ideas, such as "Within every interval of 9 numbers, there is exactly one number that is a multiple of 9." Others, whose solution attempts included "738 / 9 = 82 and 729 / 9 = 81," realized for the first time that when two adjacent multiples of 9 (e.g., 738 and 729) are divided by 9, the two quotients that are obtained (i.e., 82 and 81) are consecutive.

The examples of students' work that were presented in the previous section illustrated what we might call "variants" of the division algorithm. For example, the techniques of some students evolved to take the form of carrying out a trial division by 9, followed by the multiplication of the truncated quotient with 9 in order to see how far the product was from the initial number—an approach that we have named the "division algorithm invoking trial division" (e.g., 989 / 9 = 109.8888889, 9 × 109 = 981, 989 − 8 = 981, 981 / 9 = 109, and so on). A less sophisticated variant of this technique involved *looking at the size of the decimal portion* to provide a clue about how close the given number was to a multiple of 9.

Another variation on the division algorithm—one where the element of division was not explicit—was based on what we have named the "division algorithm invoking trial multiplication." This approach involved carrying out perhaps several trial multiplications in order to find the value of c, as in, for example, the implicit relation, $989 = 9 \times c + d$ (e.g., 9 × 106 = 954, 9 × 108 = 972, 9 × 109 = 981—the latter trial clearly bringing the solver into the interval that is within 9 on either side of the given number 989).

Whereas the searches above for multiples of 9 involved two factors, the technique developed by Nicolas involved three factors. Even though his technique did not involve prime factors, it did remind us of the *fundamental theorem of arithmetic* (i.e., "Every integer $n \geq 2$ is either a prime or a product of primes, and the product is unique apart from the order in which the factors appear") because it involved an attempt to express the given number, or its close neighbors, as a product of the factors that were acceptable according to the rules of the game (i.e., the factors from 2 to 9). As we have seen above (in fig. 3.10), he began with the product of 9 × 9 and, for given numbers less than 9 × 9 × 9, proceeded to work with potential third factors that he adjusted, along with the second factor, until he arrived at a product that was within 9 of the given number. His technique, which we named "trial multiplication involving three factors" and synthesized as a = b × c × d, where b, c, and d are whole numbers between 2 and 9 and a ≤ 729, would obviously not work for all of the numbers encountered during the week's activities, but it nevertheless reflected an important evolution in the development of his number sense.

The students who developed the techniques above seemed motivated by a need to better control their search processes. This control evolved from techniques that were initially founded on much trial and error in the successive dividing of the given numbers by 9, then 8, then 7, then 6, and so on, in the search for a quotient that was a whole number. That the evolution occurred speaks powerfully for the initial use of trial-and-error methods that sparked the genesis of stronger and more controlled techniques. However, had it not been for the presence of the computing technology, the tasks that led to the development of such techniques, although doable, would surely not have been feasible.

Concluding Remarks

The preceding discussion points to the impossibility of separating the three components of technique, task, and theory. The techniques that emerged among the students were a response to the tasks—tasks that were designed to take full advantage of the numeric capabilities of the accompanying technological tool. But more important, these techniques, which evolved, led to the growth of new mathematical knowledge for the students. That is, in generating new technology-supported techniques to solve more adequately the tasks at hand, students developed new ways of thinking about the mathematics of the tasks. Thus, as Lagrange (2000) has argued, the technological tools allowed students to "develop new techniques that constitute a bridge between tasks and theories" (p. 17).

In the elementary and middle school mathematics classroom, the calculator is often used merely as a tool for checking answers. From this study, we have been able to show the power of calculating technology for generating mathematical knowledge. The findings of the research described above suggest that with numerical-technological environments similar to the one we designed and with sustained use of the tool over a certain period of time, students will develop mathematical theory in their interaction with the technology.

REFERENCES

Artigue, Michèle. "Learning Mathematics in a CAS Environment: The Genesis of a Reflection about Instrumentation and the Dialectics between Technical and Conceptual Work." *International Journal of Computers for Mathematical Learning* 7 (2002): 245–74.

Chevallard, Yves. "L'Analyse des Pratiques Enseignantes en Théorie Anthropologique du Didactique." *Recherches en Didactique des Mathématiques* 19 (1999): 221–66.

Guin, Dominique, and Luc Trouche. "The Complex Process of Converting Tools into Mathematical Instruments: The Case of Calculators." *International Journal of Computers for Mathematical Learning* 3 (1999): 195–227.

Lagrange, Jean-Baptiste. "L'Intégration d'Instruments Informatiques dans l'Enseignement: Une Approche par les Techniques." *Educational Studies in Mathematics* 43 (2000): 1–30.

National Council of Teachers of Mathematics (NCTM). *Principles and Standards for School Mathematics.* Reston, Va.: NCTM, 2000.

Vergnaud, Gérard. "Multiplicative Structures." In *Number Concepts and Operations in the Middle Grades,* edited by James Hiebert and Merlyn Behr, pp. 141–61. Vol. 2, Research Agenda for Mathematics Education. Reston, Va.: National Council of Teachers of Mathematics, 1988.

Verillon, Pierre, and Pierre Rabardel. "Cognition and Artifacts: A Contribution to the Study of Thought in Relation to Instrumented Activity." *European Journal of Psychology of Education* 10 (1995): 77–101.

Williams, Doug, and Max Stephens. "Activity 1: Five Steps to Zero." In *Calculators in Mathematics Education,* 1992 Yearbook of the National Council of Teachers of Mathematics (NCTM), edited by James T. Fey, pp. 233–34. Reston, Va.: NCTM, 1992.

Zazkis, Rina, and Stephen Campbell. "Divisibility and Multiplicative Structure of Natural Numbers: Preservice Teachers' Understanding." *Journal for Research in Mathematics Education* 27 (November 1996): 540–63.

4

Young Children and Technology: What's Appropriate?

Douglas H. Clements
Julie Sarama

ABOUT a decade ago, we argued that "we no longer need to ask whether the use of technology is 'appropriate'" in early childhood and primary education (Clements and Swaminathan 1995). For example, we reviewed research that showed that young children are comfortable and confident in using computer software. They can understand, think about, and learn from their computer activity (Clements and Nastasi 1993). We were surprised and disappointed to find that more recent publications still warned that computers should be rejected because they pose hazards to children intellectually, developmentally, and emotionally (Cordes and Miller 2000). Unfortunately, these publications ignored or misinterpreted most of the research (Clements and Sarama 2003). So it is important to ask, What *does* the research say about what we *should* and *should not* do with technology in early childhood and primary-grade classrooms?

In this article, we present a concise summary of research implications. We begin by describing what influence technology might have on young children's learning of mathematics. We briefly summarize research on other important issues, such as social and emotional development. We then outline areas of agreement between critics and proponents of technology use. Finally, in the longest section of this paper, we draw *educational implications* from all this research.

This paper was based on work supported in part by the National Science Foundation under Grant No. ESI-9730804, "Building Blocks—Foundations for Mathematical Thinking, Pre-Kindergarten to Grade 2: Research-based Materials Development," and by the Interagency Educational Research Initiative (NSF, DOE, and NICHHD) Grant No. 0228440, "Scaling Up the Implementation of a Pre-Kindergarten Mathematics Curricula: Teaching for Understanding with Trajectories and Technologies." Any opinions, findings, and conclusions or recommendations expressed in this material are those of the author(s) and do not necessarily reflect the views of the funding agencies.

Editor's note: The CD accompanying this yearbook contains hyperlinked URLs for the Web sites that are relevant to this article.

Learning Mathematics

How effective is technology in helping young children learn mathematics? The short answer is that it is no panacea, but it can make a significant contribution. The nature and extent of its contribution depends largely on what type of technology we use.

Computer Assisted Instruction (CAI)

Children can use CAI, in which the computer presents information or tasks and gives feedback, to practice their skills. For example, drill-and-practice software can help young children develop competence in such skills as counting and sorting (Clements and Nastasi 1993). Indeed, some reviewers claim that the largest gains in the use of CAI have been in mathematics for preschool (Fletcher-Flinn and Gravatt 1995) or primary-grade children, especially in compensatory education (Lavin and Sanders 1983; Niemiec and Walberg 1984; Ragosta, Holland, and Jamison 1981). About ten minutes a day proved sufficient time for significant gains; twenty minutes was even better. This CAI approach may be as or more cost effective than traditional instruction (Fletcher, Hawley, and Piele 1990) and other instructional interventions, such as peer tutoring and reducing class size (Niemiec and Walberg 1987).

Properly chosen, computer games that provide drill may also be effective. Kraus (1981) reported that second graders with an average of one hour of interaction with a computer game over a two-week period responded correctly to twice as many items on an addition facts speed test as children in a control group.

What aged children can benefit? An old concern is that children must reach the stage of concrete operations before they are ready to work with computers. Recent research, however, has found that preschoolers are more competent than has been thought and exhibit thinking traditionally considered "concrete" (Gelman and Baillargeon 1983). Moreover, what is concrete to the child has more to do with what is meaningful and manipulable than with physical characteristics, as we shall see. However, people still ask, Should young children use computers at all? Does this "rush" them? Research indicates that anything, books or pencils, can be used inappropriately, but that all media, computers included, can provide experiences that facilitate children's learning in many areas in ways consistent with children's development. In sum, children as young as preschool age can benefit socially, emotionally, and cognitively from using computers. For example, three-year-olds learn sorting from a computer task as easily as from a concrete doll task (Brinkley and Watson 1987–88), and kindergartners learn counting as well or better with a computer (Hungate 1982).

However successful, drill should be used in moderation. Some children may be less motivated to perform academic work or less creative following drill (Clements and Nastasi 1985; Haugland 1992), and their creativity may be harmed by a consistent diet of drill. Also, using only drill software would fall short of the vision of the National Council of Teachers of Mathematics (NCTM) that children should be mathematically literate (2000). Other approaches help achieve that vision. In the next two sections, we describe two such approaches. In one, children manipulate mathematical objects on the computer screen. In another, children give directions to the computer to create geometric shapes. There are many new developments, but these two have been well studied.

Computer Manipulatives

Most of us think of "manipulatives" as physical objects. However, a somewhat surprising research finding is that manipulating shapes and objects on the computer can be just as or *more* effective in supporting learning (Clements and McMillen 1996). For example, in one study, the first time children reflected on, and planned, putting together shapes to make new shapes, they were working on a computer, not with physical blocks (Sarama, Clements, and Vukelic 1996). In a similar vein, children who explore shapes on the computer learn to understand and apply concepts such as symmetry, patterns, and spatial order (Wright 1994), often surpassing their teachers' expectations. In a study comparing the use of physical bean sticks and on-screen bean sticks, children found the computer manipulative easier to use for learning (Char 1989).

One of the reasons for this finding is that computer manipulatives allow children to perform specific mathematical transformations on objects on the screen. For example, whereas physical base-ten blocks must be "traded" (when subtracting, children may need to trade 1 ten for 10 ones), children can break a computer base-ten block directly into 10 ones (see figs. 4.1 and 4.2). Such actions are more in line with the *mental actions* that we want children to carry out. The computer also *connects* the blocks to the symbols. For example, the number represented by the base-ten blocks is dynamically connected to the children's actions on the blocks, so that when the child changes the blocks, the number displayed is automatically changed as well. In figure 4.3, as the child removes 4 tens from the 320 previously shown in figure 4.2, the bottom row automatically adjusts. Such features can help children make sense of their activity, the numbers, and the arithmetic.

In summary, computer manipulatives can offer unique advantages (Clements and Sarama 1998; Sarama, Clements, and Vukelic 1996). They can

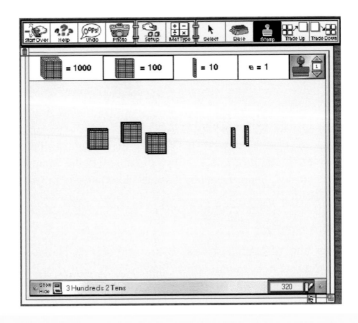

Fig. 4.1. The base-ten blocks from *Math Tool Chest*, from McGraw-Hill. The child moves blocks onto the work area, and the numerals automatically update.

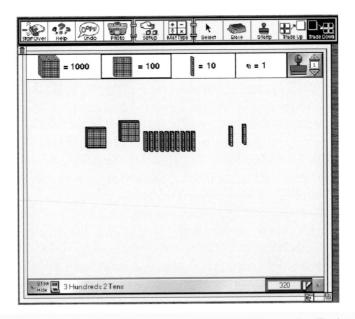

Fig. 4.2. Wishing to subtract 40 from 320, the child uses the "Trade Down" feature to break apart one of the hundreds blocks.

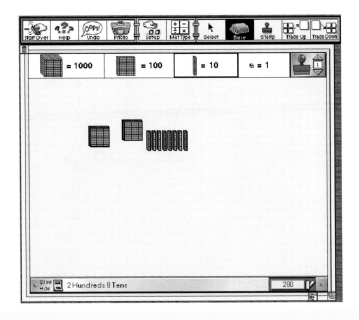

Fig. 4.3. As the child removes 4 tens, the numerals and number words again update automatically. Note that if teachers or students want students to "check themselves," the numerals and number words can be hidden with the switches located next to the displays.

allow children to save and retrieve work (and that work doesn't get "bumped" and "ruined"!) and thus work on projects over a long period (Ishigaki, Chiba, and Matsuda 1996). Computers can offer a flexible and manageable manipulative, one that, for example, might "snap" into position. They can provide an extensible manipulative, which children can resize or cut. For example, in "Create a Scene," one of the programs from the Building Blocks project, children can turn, flip, resize, glue, and even cut shapes (see fig. 4.4, where a child is making two new triangles by cutting a trapezoid). Computer manipulatives can also help connect concrete and symbolic representations by means of multiple, linked representations and feedback, such as showing base-ten blocks dynamically linked to numerals (figs. 4.1–4.3). Computers can record and replay children's actions, encouraging their reflection. In a similar vein, computers can help bring mathematics to explicit awareness, by asking children consciously to choose what mathematical operations (turn, flip, scale) to apply. For example, when children flip or turn shapes in Create a Scene (fig. 4.4), our research shows that they consciously think and talk about the geometric motions they are using, which they did not do with physical manipulatives (Sarama, Clements, and Vukelic 1996).

Fig. 4.4. The DLM Early Childhood Math Software (Clements and Sarama 2003), from the Building Blocks project (www.gse.buffalo.edu/org/buildingblocks/), includes the activity "Create a Scene."

Finally, technology may also foster deeper conceptual thinking, including a valuable type of "cognitive play" (Steffe and Wiegel 1994). For example, to develop concepts of length and measurement, children engaged in drawing on-screen sticks, changing the color of a stick, marking a stick, breaking it along the marks, joining the parts back together, and cutting off pieces from a stick. Children adopted a playful attitude as they repeatedly engaged in these activities, and they learned considerable mathematics.

Piaget, Papert, and Turtle Geometry

Geometric and spatial relationships can and should be learned through bodily movement (Piaget and Inhelder 1967). Just as computer manipulatives can extend children's experiences with feeling and manipulating shapes, on-screen "turtle geometry" can extend their experiences such as walking around a rectangular rug or around the playground. Seymour Papert (1980) invented the computer turtle for this purpose. After walking around the rug, children might direct an on-screen "turtle" to "walk" (or form) a rectangle with commands such as "forward 100 right 90 (degrees), forward 50 ..." (fig. 4.5). Why not

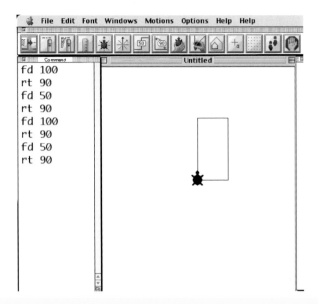

Fig. 4.5. Two primary-grade children make a rectangle with the turtle with *Turtle Math* (Clements and Meredith 1994; Clements and Sarama 1996).

just draw on paper? First, drawing a geometric shape on paper, for example, is for most people a motor procedure. In creating a Logo procedure to draw the shape, however, children must analyze the visual aspects of the shape and their movements in drawing it, thus requiring them to reflect on how the parts of shapes are put together. Writing a sequence of Logo commands, or a procedure, to draw a shape "allows, or obliges, the child to externalize intuitive expectations. When the intuition is translated into a program it becomes more obtrusive and more accessible to reflection" (Papert 1980, p. 145).

That *does* happen. Primary-grade children show greater explicit awareness of the properties of shapes and the meaning of measurements after working with the turtle (Clements and Nastasi 1993). They learn about the measurement of length (Campbell 1987; Clements et al. 1997; Sarama 1995) and angle (Browning 1991; Clements and Battista 1989; du Boulay 1986; Kieran and Hillel 1990; Olive, Lankenau, and Scally 1986). For example, children working with the turtle gradually replace full rotations of their bodies with smaller rotations of an arm, hand, or finger and eventually internalize these actions as mental imagery. They combine their ideas of turning their bodies and turning as a number (e.g., "right 90") into a meaningful concept of turn and angle measure (Clements et al. 1996; Clements and Burns 2000).

Logo is not easy to learn or use. However, as one primary-grade child declared, "This picture was very hard and it took me 1 hour and 20 minutes to do it, but it had to be done. I liked doing it" (Carmichael et al. 1985, p. 90). Children work hard to create combinations of mathematical objects and ideas to make designs (e.g., fig. 4.6). Moreover, when the environment is gradually and systematically introduced to the children and when the interface is age-appropriate, even young children learn to control the turtle and learn mathematics (Allen, Watson, and Howard 1993; Clements 1983–84; Howard, Watson, and Allen 1993; Stone 1996; Watson, Lange, and Brinkley 1992). A study of 1624 children and their teachers found that children using a Logo-based geometry curriculum made about *double* the gains of the control groups in mathematics concepts, reasoning, and problem solving (Clements, Battista, and Sarama 2001). These studies and hundreds of others (Clements and Sarama 1997) indicate that Logo, *used thoughtfully,* can provide an additional evocative context for young children's explorations of mathematical ideas. Such "thoughtful use" includes structuring and guiding Logo work to help children form strong, valid mathematical ideas (Clements, Battista, and Sarama 2001). Children often do not appreciate the mathematics in Logo work unless someone helps them see the work mathematically. Effective teachers

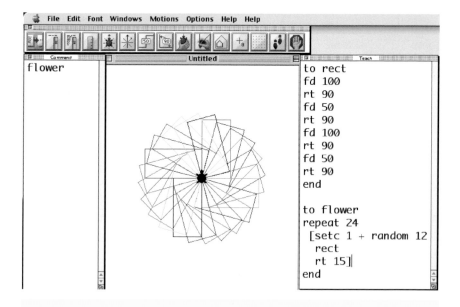

Fig. 4.6. The children *use* the rectangle procedure they made, along with other commands, to make flower petals.

raise questions about "surprises" or conflicts between children's intuitions and computer feedback to promote children's reflection. They pose challenges and tasks designed to make the mathematical ideas explicit for children. They help children build bridges between the Logo experience and their regular mathematics work (Clements 1987; Watson and Brinkley 1990/91).

Social and Emotional Development

An old concern is that educational computer programs will isolate children. Research and practice indicate just the opposite—*computers can serve as catalysts for positive social interaction and emotional growth.* As one example, children spent nine times as much time talking to peers while on the computer than while doing puzzles (Muller and Perlmutter 1985). Other research shows that the nature of their interaction tends to be positive. For example, *children overwhelmingly display positive emotions when using computers* (Shade 1994), *especially when they use computers together* (Ishigaki, Chiba, and Matsuda 1996). Similarly, *computer work can instigate new instances and forms of collaborative work such as helping or teaching, discussing and building on one another's ideas, cooperating, and praising*—and *social interaction increases between children with disabilities and their normally developing peers* (Clements and Sarama 2003; Hutinger and Johanson 2000).

Finally, *computers can facilitate both social and cognitive interactions— each to the benefit of the other.* Good software encourages children to talk about their work as well as engage in more advanced cognitive type of play than they do in other centers (Genishi, McCollum, and Strand 1985; Hoover and Austin 1986). Indeed, one study found that only the computer center engendered high levels of both language development and cooperative play (Muhlstein and Croft 1986).

Points on Which Critics and Advocates Agree

Both critics and advocates of computers agree in several areas (Clements and Sarama 2003, Sarama and Clements 2002). These are likely to be important points for educators to keep in mind.

We should advocate meaningful, "whole" development and learning for children. The literature supports meaningful, holistic development and learning. Although some critics contend this excludes computer work, the same approach underlies many curriculum and software developers' efforts to create computer environments that promote understanding, child-centered learning, and connections among experiences and areas of knowledge.

Technology can help children with disabilities. Whether providing instruction or adaptive devices, technology offers important benefits to children with disabilities. Software can provide individualized instruction with patient feedback, as well as opportunities to explore mathematics. Adaptive devices can facilitate communication, movement, and control of the environment. Specific software can help teachers work with and track children's progress on IEPs. Children in comprehensive, technology-enhanced programs make progress in all developmental areas, including social-emotional, fine motor, gross motor, communication, cognition, and self-help skills (Hutinger et al. 1998; Hutinger and Johanson 2000). It is unsurprising, then, that both critics and proponents of technology support such use. We argue that the similarities of children with and without disabilities indicate that *all* children can benefit from high-quality use of technology, and all are harmed by inappropriate use.

There are inappropriate uses of computers. Advocates agree with critics that some applications of technology may be detrimental. For example, we have argued against too heavy an emphasis on drill; and computer management that replaces decision making by teachers and children represents an emphasis on bureaucratic efficiency and a de-emphasis on young children's conceptual development (Clements, Nastasi, and Swaminathan 1993). In a similar vein, the discussion of the Technology Principle in *Principles and Standards for School Mathematics* (NCTM 2000) states, "Technology should not be used as a replacement for basic understandings and intuitions; rather, it can and should be used to foster those understandings and intuitions" (p. 25). We have also raised concerns about aggression and inequities (Clements 1985), and there is every reason to limit exposure of all children to violence in all media, especially those at risk developmentally and socially.

What about the Internet? We believe that there is reason both for concern about inappropriate content (because many parents are not placing any restriction on Internet use, according to Shields and Behrmann [2000]) and for potential benefit. We believe that more research on *specific uses* is needed before we make firm recommendations.

There are inappropriate justifications for early computer use. Most advocates and critics agree that primary-grade children do not need technology training in preparation for future employment. The use of computers with children is more about realizing their potential across the many critical areas of development in the young child, including the intellectual and the social-emotional domains. When computers contribute to this development, they should be used. When they do not, they should not be used.

The total time in front of screens should be limited. The time children spend in front of a screen should be limited. Screen time is dominated by commercial TV and often inappropriate video games, with children aged two to seven spending from two to three hours a day in front of a screen (Subrahmanyam et al. 2000). These should be limited first (Clements and Nastasi 1993; Shields and Behrmann 2000). More positive uses, such as those suggested here, also should be limited, to about one to two hours a day (American Academy of Pediatrics 1999). However, strict time limits in an early childhood classroom may not be wise. For example, setting a time for five or ten minutes for each child may generate hostility and isolation. It also may keep children from communicating and sharing (Hutinger et al. 1998). Research shows that giving children some say about how long they need to complete their computer work may be a more positive strategy.

Even high-quality technology is not always used. Everyone might agree that a certain use of technology is beneficial ... and it may often remain on the shelf. It is essential to have adequate professional development, technical and instructional support, and cooperation among all parties. This leads to our final point of agreement.

Money is often spent unwisely. Money should not be spent without reflection on educational priorities. If high-priority goals can be achieved well with specific computer applications, then detailed planning, which includes long-range plans for professional development and support, should precede any purchase.

Strategies for Effective Teaching with Technology

Research offers numerous suggestions for effective teaching with technology. In this section, we draw implications from research regarding arranging and managing computers in the classroom, choosing software, strategies for interacting with children in computer environments, and supporting children who have special needs.

Arranging Computers in the Classroom

Initial adult support helps young children use computers to learn (Rosengren et al. 1985; Shade et al. 1986). With such help, they can often use computers independently. Still, children are more attentive, more engaged, and less frustrated when an adult is at least nearby (Binder and Ledger 1985). So, make the computer one of many choices, placed where you or other adults can supervise and assist children (Sarama and Clements 2002).

You can enhance the social use of computers by planning the physical arrangement of computers and furniture carefully (Davidson and Wright 1994; Shade 1994). Place the parts of the computer with which the children interact—the keyboard, mouse, monitor, and microphone—at the children's eye level, on a low table or even on the floor. If children are changing CD-

ROMs, place them so that children can see and change them easily. Place other parts of computer equipment out of children's reach. Stabilize and lock down all parts as necessary. If computers are to be shared with other classrooms, consider using stable rolling carts.

Place two seats in front of the computer and one at the side for an adult to encourage positive social interaction. If more than two very young children work with a computer simultaneously, they assert the right to control the keyboard frequently (Shrock et al. 1985). Instead, consider having one child on a computer but placing two computers close to each other to facilitate the sharing of ideas. Primary-grade children can learn to take turns, if those turns are well defined. Cooperative use of computers for those who can collaborate raises achievement (Xin 1999); a mixture of use by pairs and by individuals may be ideal (Shade 1994).

Locate computers centrally in the classroom to invite other children to pause and participate in the computer activity. Such an arrangement also helps keep adult participation at an optimum level. That is, adults are nearby to provide supervision and assistance as needed (Clements 1991). Other factors, such as the ratio of computers to children, may also influence social behaviors. Establish less than a 10:1 ratio of children to computers to encourage computer use, cooperation, and equal access to girls and boys (Lipinski et al. 1986; Yost 1998).

To encourage children to connect off-computer and on-computer experiences, place print materials, manipulatives, and real objects next to the computer (Hutinger and Johanson 2000). This arrangement also furnishes good activities for children who are observing or waiting for their turn. Change software, along with corresponding nontechnology centers, to match your educational themes.

Managing the Computer Environment

As with any center, teach children proper computer use and care, and post signs to remind them of the rules (e.g., no liquids, sand, food, or magnets near computers). Use a child-oriented software utility to help children find and use the programs they want. This practice also prevents them from inadvertently harming other programs or files and makes everyone's life easier.

Monitor the time that children spend on computers, and make sure that everyone is given fair access. Note, though, that at least one study has found that rigid time limits generated hostility and isolation instead of social communication (Hutinger and Johanson 2000). Instead, use flexible times with sign-up lists that encourage children to manage themselves. The sign-up list itself has a positive effect on preschoolers' emergent literacy (Hutinger and Johanson 2000).

Introduce computer work gradually. Initially, use only one or two programs at a time. Expect independent work from children gradually. Prepare them for independence, even having individuals or small groups of children work closely with an adult at first, and then increase the degree of such work slowly. Offer substantial support and guidance initially, even sitting with children at the computer to encourage taking turns. Then gradually foster self-directed and cooperative learning.

When necessary, teach children effective collaboration; for example, communication and negotiation skills. For young children, this might include such matters as what constitutes a "turn" in a particular game. However, do not mandate sharing the computer all the time. Especially with construction-oriented programs, children sometimes need to work alone. If possible, as recommended previously, make at least two computers available so that peer teaching and other kinds of interaction can take place, even if children are working on one computer.

After children are working independently, offer enough guidance, but not too much. Intervening too much or at the wrong times can decrease peer tutoring and collaboration (Bergin, Ford, and Mayer-Gaub 1986; Emihovich and Miller 1988; Riel 1985). On the contrary, without any adult guidance, children tend to "jockey" for position at the computer and use it in the turn-taking, competitive manner of video games (Lipinski et al. 1986; Silvern, Countermine, and Williamson 1988).

Research shows that the introduction of a microcomputer often places many additional demands on the teacher (Shrock et al. 1985). Plan carefully so that you have adequate adult assistance and plan to use only technology that will substantially benefit your students.

Choosing Software

Using computers positively in these ways assumes that the teacher has chosen high-quality software. Previous sections already described the nature of such software; here, we stress that software should always be selected according to sound educational principles. Preschool children respond correctly more often using software that incorporates such principles (Grover 1986; see also Haugland and Shade 1990; Haugland and Wright 1997), such as the following.

Actions and graphics provide a meaningful context for children. Children should be able to make sense of the activities, both the situation or task and the mathematics.

The reading level, assumed attention span, and means of responding should be appropriate for the age level. Instructions should be clear, including, for example, simple choices in the form of a picture menu.

After enjoying initial adult support, children should be able to use the software independently. There should also be multiple opportunities for success.

Feedback is informative. Feedback should not be just "right" or "wrong."

Children should be in control. Software should provide as much manipulative power as possible.

Software allows children to create, program, or invent new activities. It has the potential for independent use but also challenges children. It is flexible and allows more than one correct response.

Software enhances the classroom environment. Computer programs should help and empower children to learn and meet specific educational and developmental goals more effectively and powerfully than they could without the technology. This point serves as an important summary: The computer should not be simply an end unto itself. Computers can help children learn and should be used reflectively by both children and their teachers. Children should learn to understand how and why the programs they use work the way they do (Turkle 1997).

Effective Strategies for Teaching with Computers

Essential to the effective use of computers is teacher planning, participation, and support. Your role should be that of a facilitator of children's learning. Such facilitation includes not only physical structuring of the environment but also establishing standards for, and supporting specific types of, learning environments. When using open-ended programs, for example, provide considerable support to build children's independent use. Structure and discuss computer work to help children form viable concepts and strategies, posing questions to help them reflect on these concepts and strategies, and "building bridges" to help children connect their computer and noncomputer experiences.

Stay active. Closely guide children's learning of basic tasks, and then encourage experimentation with open-ended problems. Encourage, question, prompt, and demonstrate, *without* offering unnecessary help or limiting children's opportunity to explore (Hutinger and Johanson 2000). Redirect inappropriate behaviors, model strategies, and give children choices (Hutinger et al. 1998). Such scaffolding leads children to reflect on their own thinking behaviors and brings higher-order thinking processes to the fore. Use metacognitively oriented instruction, including strategies such as identifying goals, active monitoring, modeling, questioning, reflecting, peer tutoring, discussion, and reasoning (Elliott and Hall 1997).

Make the subject matter to be learned clear and extend the ideas children encounter. Focus attention on the crucial aspects and ideas of the activities.

When appropriate, facilitate disequilibrium and its resolution by using the computer feedback to help children reflect on and question their ideas and eventually strengthen their concepts. Help children build links between computer and noncomputer work.

Frequently lead whole-group discussions that help children communicate about their solution strategies and reflect on what they have learned (Galen and Buter 2000). Avoid using overly directive teaching behaviors (except as necessary and on topics such as using the computer equipment) and, as stated earlier, setting strict time limits. Instead, prompt children to teach each other by physically placing one child in a teaching role or verbally reminding a child to explain his or her actions and respond to specific requests for help (Paris and Morris 1985).

Remember to provide scaffolding consistently. In one study (Yelland 1998), children were only given instructions for specific tasks and then mostly left alone. These children rarely planned, were often off task, rarely cooperated, displayed frustration and lack of confidence, and did not finish tasks. In another study (Yelland 1994), the teacher scaffolded instruction by presenting open-ended but structured tasks, holding group brainstorming sessions about problem-solving strategies, encouraging children to work collaboratively, asking them to think and discuss their plans before working at the computer, questioning them about their plans and strategies, and modeling strategies as necessary. These children planned, worked on tasks collaboratively, were able to explain their strategies, were rarely frustrated, and completed tasks efficiently. They showed a high level of mathematical reasoning about geometric figures and motions as well as number and measurement.

In using open-ended, constructive software, seek a balance between teacher guidance and children's self-directed exploration (Escobedo and Bhargava 1991). Give children designated projects instead of merely suggesting that they "free explore" (Lemerise 1993), and model and share projects (Hall and Hooper 1993). They will work for a longer time and actively search for diverse ways to solve the task. Children who just free explore quickly lose interest.

With all programs, but especially with open-ended software, structure and guide work to help children form strong, valid mathematical ideas (Clements, Battista, and Sarama 2001). Children often do not appreciate the mathematics in such work unless someone helps them see the work mathematically. Pose challenges and tasks designed to make the mathematical ideas explicit for children. Help children build bridges between their computer experiences and their regular mathematics work (Clements 1987; Watson and Brinkley 1990/91).

Remember that preparation and follow-up are as necessary for computer activities as they are for any other kinds of activities. Do not omit essential whole-group discussion sessions following computer work. Consider using a single computer with a large screen or with projection equipment.

Supporting Children with Special Needs

Make sure that special education children are accepted, supported, and included in regular classroom work (Xin 1999). Choose software that matches your educational goals for the children and spend the time to guide children to use it successfully. Choose computer-assisted instruction (CAI) that appears patient and nonjudgmental, proceeds at the child's pace, and provides immediate reinforcement (Schery and O'Connor 1997). Also, use exploratory and problem-solving software (Clements and Nastasi 1988; Lehrer et al. 1986; Nastasi, Clements, and Battista 1990). Finally, follow the Division for Early Childhood (Council for Exceptional Children) recommended practices for technology applications (Sandall, McLean, and Smith 2000). Use CAI for a wide variety of purposes: to increase communication and develop language skills, environmental access, social-adaptive skills, mobility and orientation skills, daily life skills, social interaction skills, and health awareness. Also use technology for instruction, assessment, and as a way to communicate and coordinate with families.

Conclusion

In the early childhood and primary grades, as much as in later grades, technology can enhance and support the vision of mathematics teaching and learning offered in *Principles and Standards of School Mathematics* (NCTM 2000). Young children understand and benefit from appropriate computer activities. Computer-assisted instruction, such as drill and practice and games, can help children develop mathematical skills. Other approaches, such as using computer manipulatives and tools that emphasize problem solving, offer unique advantages and have the potential to facilitate deeper conceptual thinking. These cognitive gains are complemented by benefits for children's social and emotional development. Critics and advocates alike have research-based "common grounds" for agreement that should allow everyone to move forward with technology use. Finally, implications from this body of research provide specific guidelines for the effective and appropriate use of technology. As the discussion of NCTM's Technology Principle states, "Students can learn more mathematics more deeply with the appropriate use of technology" (NCTM 2000, p. 25).

References

Allen, Jean, J. Allen Watson, and Janice R. Howard. "The Impact of Cognitive Styles on the Problem Solving Strategies Used by Preschool Minority Children in Logo Microworlds." *Journal of Computing in Childhood Education* 4 (1993): 203–17.

American Academy of Pediatrics. "Media Education." *Pediatrics* 104 (1999): 341–43.

Bergin, David A., Martin E. Ford, and Gabrael Mayer-Gaub. "Social and Motivational Consequences of Microcomputer Use in Kindergarten." San Francisco: American Educational Research Association, 1986.

Binder, Sari-Lynn, and Barbara Ledger. *Preschool Computer Project Report.* Oakville, Ont.: Sheridan College, 1985.

Brinkley, Vickie M., and J. Allen Watson. "Effects of Microworld Training Experience on Sorting Tasks by Young Children." *Journal of Educational Technology Systems* 16 (1987–88): 349–64.

Browning, Christine A. "Reflections on Using Lego® Tc Logo in an Elementary Classroom." In *Proceedings of the Third European Logo Conference,* edited by Eduardo Calabrese, pp. 173–85. Parma, Italy: Associazione Scuola e Informatica, 1991.

Campbell, Patricia F. "Measuring Distance: Children's Use of Number and Unit." Final report submitted to the National Institute of Mental Health under the Adamha Small Grant Award Program Grant No. Msma 1 R03 Mh423435-01. College Park, Md.: University of Maryland, 1987.

Carmichael, Hilda W., J. Dale Burnett, William C. Higginson, Barbara G. Moore, and Phyllis J. Pollard. *Computers, Children and Classrooms: A Multisite Evaluation of the Creative Use of Microcomputers by Elementary School Children.* Toronto, Ont.: Ministry of Education, 1985.

Char, Cynthia A. "Computer Graphic Feltboards: New Software Approaches for Young Children's Mathematical Exploration." San Francisco: American Educational Research Association, 1989.

Clements, Douglas H. "Supporting Young Children's Logo Programming." *Computing Teacher* 11, no. 5 (1983–84): 24–30.

———. "Technological Advances and the Young Child: Television and Computers." In *Young Children in Context: Impact of Self, Family and Society on Development,* edited by Caven S. Mcloughlin and Dominic F. Gullo, pp. 218–53. Springfield, Ill.: Charles Thomas, 1985.

———. "Longitudinal Study of the Effects of Logo Programming on Cognitive Abilities and Achievement." *Journal of Educational Computing Research* 3 (1987): 73–94.

———. "Current Technology and the Early Childhood Curriculum." In *Yearbook in Early Childhood Education,* Vol. 2: *Issues in Early Childhood Curriculum,* edited by Bernard Spodek and Olivia N. Saracho, pp. 106–31. New York: Teachers College Press, 1991.

Clements, Douglas H., and Michael T. Battista. "Learning of Geometric Concepts in a Logo Environment." *Journal for Research in Mathematics Education* 20 (1989): 450–67.

Clements, Douglas H., Michael T. Battista, and Julie Sarama. *Logo and Geometry, Journal for Research in Mathematics Education* Monograph No. 10. Reston, Va.: National Council of Teachers of Mathematics, 2001.

Clements, Douglas H., Michael T. Battista, Julie Sarama, and Sudha Swaminathan. "Development of Turn and Turn Measurement Concepts in a Computer-Based Instructional Unit." *Educational Studies in Mathematics* 30 (1996): 313–37.

Clements, Douglas H., Michael T. Battista, Julie Sarama, Sudha Swaminathan, and Sue McMillen. "Students' Development of Length Measurement Concepts in a Logo-Based Unit on Geometric Paths." *Journal for Research in Mathematics Education* 28 (January 1997): 70–95.

Clements, Douglas H., and Barbara A. Burns. "Students' Development of Strategies for Turn and Angle Measure." *Educational Studies in Mathematics* 41 (2000): 31-45.

Clements, Douglas H., and Sue McMillen. "Rethinking 'Concrete' Manipulatives." *Teaching Children Mathematics* 2 (January 1996): 270–79.

Clements, Douglas H., and Julie Sarama Meredith. Turtle Math. Software. Montreal, Que.: Logo Computer Systems, 1994.

Clements, Douglas H., and Bonnie K. Nastasi. "Effects of Computer Environments on Social-Emotional Development: Logo and Computer-Assisted Instruction." *Computers in the Schools* 2, no. 2–3 (1985): 11–31.

———. "Social and Cognitive Interactions in Educational Computer Environments." *American Educational Research Journal* 25 (1988): 87–106.

———. "Electronic Media and Early Childhood Education." In *Handbook of Research on the Education of Young Children,* edited by Bernard Spodek, pp. 251–75. New York: Macmillan, 1993.

Clements, Douglas H., Bonnie K. Nastasi, and Sudha Swaminathan. "Young Children and Computers: Crossroads and Directions from Research." *Young Children* 48, no. 2 (1993): 56–64.

Clements, Douglas H., and Julie Sarama. "Research on Logo: A Decade of Progress." *Computers in the Schools* 14, no. 1–2 (1997): 9–46.

———. *Building Blocks—Foundations for Mathematical Thinking, Pre-Kindergarten to Grade 2: Research-Based Materials Development.* National Science Foundation Grant No. ESI-9730804; www.gse.buffalo.edu/org/buildingblocks/. Buffalo, N.Y.: State University of New York at Buffalo, 1998.

———. "Strip Mining for Gold: Research and Policy in Educational Technology—a Response to 'Fool's Gold.'" *Educational Technology Review* 11, no. 1 (2003). Retrieved May 11, 2004, from www.aace.org/pubs/etr/issue4/clements.cfm. Clements, Douglas H., and Sudha Swaminathan. "Technology and School Change: New Lamps for Old?" *Childhood Education* 71 (1995): 275–81.

Cordes, Colleen, and Edward Miller. *Fool's Gold: A Critical Look at Computers in Childhood.* College Park, Md.: Alliance for Childhood, 2000. Retrieved November 7, 2000, from www.allianceforchildhood.net/projects/computers/computers_reports.htm.

Davidson, Jane, and June L. Wright. "The Potential of the Microcomputer in the Early Childhood Classroom." In *Young Children: Active Learners in a Technological Age,* edited by June L. Wright and Daniel D. Shade, pp. 77–91. Washington, D.C.: National Association for the Education of Young Children, 1994.

du Boulay, B. "Part II: Logo Confessions." In *Cognition and Computers: Studies in Learning,* edited by R. Lawler, B. du Boulay, M. Hughes, and H. Macleod, pp. 81–178. Chichester, England: Ellis Horwood, 1986.

Elliott, Alison, and Neil Hall. "The Impact of Self-Regulatory Teaching Strategies on 'At-Risk' Preschoolers' Mathematical Learning in a Computer-Mediated Environment." *Journal of Computing in Childhood Education* 8, no. 2/3 (1997): 187–98.

Emihovich, Catherine, and Gloria E. Miller. "Talking to the Turtle: A Discourse Analysis of Logo Instruction." *Discourse Processes* 11 (1988): 183–201.

Escobedo, Theresa H., and Ambika Bhargava. "A Study of Children's Computer-Generated Graphics." *Journal of Computing in Childhood Education* 2 (1991): 3–25.

Fletcher, J. D., David E. Hawley, and Philip K. Piele. "Costs, Effects, and Utility of Microcomputer Assisted Instruction in the Classroom." *American Educational Research Journal* 27 (1990): 783–806.

Fletcher-Flinn, Claire M., and Breon Gravatt. "The Efficacy of Computer Assisted Instruction (CAI): A Meta-Analysis." *Journal of Educational Computing Research* 12 (1995): 219–42.

Galen, Frans H. J. van, and Arlette Buter. "Computer Tasks and Classroom Discussions in Mathematics." Article published 1997 on CD-ROM by the Freudenthal Institute (Utrecht, Netherlands) and distributed to attendants at the International Congress on Mathematics Education (ICME-9), Tokyo/Makuhari, Japan, July 30–August 6, 2000.

Gelman, Rochel, and Renée Baillargeon. "A Review of Some Piagetian Concepts." In *Handbook of Child Psychology,* edited by P. H. Mussen, pp. 167–230. New York: John Wiley & Sons, 1983.

Genishi, Celia, Pam McCollum, and Elizabeth B. Strand. "Research Currents: The Interactional Richness of Children's Computer Use." *Language Arts* 62, no. 5 (1985): 526–32.

Grover, S. C. "A Field Study of the Use of Cognitive-Developmental Principles in Microcomputer Design for Young Children." *Journal of Educational Research* 79 (1986): 325–32.

Hall, Irene, and Paula Hooper. "Creating a Successful Learning Environment with Second and Third Graders, Their Parents, and Lego/Logo." In *New Paradigms in Classroom Research on Logo Learning,* edited by Daniel Lynn Watt and Molly Lynn Watt, pp. 53–63. Eugene, Ore.: International Society for Technology in Education, 1993.

Haugland, Susan W. "Effects of Computer Software on Preschool Children's Developmental Gains." *Journal of Computing in Childhood Education* 3, no. 1 (1992): 15–30.

Haugland, Susan W., and Daniel D. Shade. *Developmental Evaluations of Software for Young Children.* Albany, N.Y.: Delmar, 1990.

Haugland, Susan W., and June L. Wright. *Young Children and Technology: A World of Discovery.* Boston: Allyn & Bacon, 1997.

Hoover, Jeanne M., and Ann M. Austin. "A Comparison of Traditional Preschool and Computer Play from a Social/Cognitive Perspective." San Francisco: American Educational Research Association, 1986.

Howard, Janice R., J. Allen Watson, and Jean Allen. "Cognitive Style and the Selection of Logo Problem-Solving Strategies by Young Black Children." *Journal of Educational Computing Research* 9 (1993): 339–54.

Hungate, Harriet. "Computers in the Kindergarten." *Computing Teacher* 9 (January 1982): 15–18.

Hutinger, Patricia L., Carol Bell, Marisa Beard, Janet Bond, Joyce Johanson, and Clare Terry. "The Early Childhood Emergent Literacy Technology Research Study." Final Report. Macomb, Ill.: Western Illinois University, 1998.

Hutinger, Patricia L., and Joyce Johanson. "Implementing and Maintaining an Effective Early Childhood Comprehensive Technology System." *Topics in Early Childhood Special Education* 20, no. 3 (2000): 159–73.

Ishigaki, Emiko Hannah, Takeo Chiba, and Sohei Matsuda. "Young Children's Communication and Self Expression in the Technological Era." *Early Childhood Development and Care* 119 (1996): 101–17.

Kieran, Carolyn, and Joel Hillel. "'It's Tough When You Have to Make the Triangles Angles'": Insights from a Computer-Based Geometry Environment." *Journal of Mathematical Behavior* 9 (1990): 99–127.

Kraus, William H. "Using a Computer Game to Reinforce Skills in Addition Basic Facts in Second Grade." *Journal for Research in Mathematics Education* 12 (March 1981): 152–55.

Lavin, Richard J., and Jean E. Sanders. "Longitudinal Evaluation of the C/A/I Computer Assisted Instruction Title 1 Project: 1979–82." Chelmsford, Mass.: Merrimack Education Center, 1983.

Lehrer, Richard, Laura D. Harckham, Philip Archer, and Robert M. Pruzek. "Microcomputer-Based Instruction in Special Education." *Journal of Educational Computing Research* 2 (1986): 337–55.

Lemerise, Tamara. "Piaget, Vygotsky, & Logo." *Computing Teacher* 20 (April 1993): 24–28.

Lipinski, Judith M., Robert E. Nida, Daniel D. Shade, and J. Allen Watson. "The Effects of Microcomputers on Young Children: An Examination of Free-Play Choices, Sex Differences, and Social Interactions." *Journal of Educational Computing Research* 2 (1986): 147–68.

Muhlstein, Eleanor A., and Doreen J. Croft. "Using the Microcomputer to Enhance Language Experiences and the Development of Cooperative Play among Preschool Children." Cupertino, Calif.: De Anza College, 1986. (ERIC Document Reproduction Service No. ED269004)

Muller, Alexandra A., and Marion Perlmutter. "Preschool Children's Problem-Solving Interactions at Computers and Jigsaw Puzzles." *Journal of Applied Developmental Psychology* 6 (1985): 173–86.

Nastasi, Bonnie K., Douglas H. Clements, and Michael T. Battista. "Social-Cognitive Interactions, Motivation, and Cognitive Growth in Logo Programming and CAI Problem-Solving Environments." *Journal of Educational Psychology* 82 (1990): 150–58.

National Council of Teachers of Mathematics (NCTM). *Principles and Standards for School Mathematics.* Reston, Va.: NCTM, 2000.

Niemiec, Richard P., and Herbert J. Walberg. "Computers and Achievement in the Elementary Schools." *Journal of Educational Computing Research* 1 (1984): 435–40.

———. "Comparative Effects of Computer-Assisted Instruction: A Synthesis of Reviews." *Journal of Educational Computing Research* 3 (1987): 19–37.

Olive, John, C. A. Lankenau, and Susan P. Scally. "Teaching and Understanding Geometric Relationships through Logo: Phase II. Interim Report: The Atlanta-Emory Logo Project." Atlanta, Ga.: Emory University, 1986.

Papert, Seymour. *Mindstorms: Children, Computers, and Powerful Ideas.* New York: Basic Books, 1980.

Paris, Cynthia L., and Sandra K. Morris. "The Computer in the Early Childhood Classroom: Peer Helping and Peer Teaching." Paper presented at the Microworld for Young Children Conference, College Park, Md., 1985.

Piaget, Jean, and Bärbel Inhelder. *The Child's Conception of Space.* Translated by F. J. Langdon and J. L. Lunzer. New York: W. W. Norton, 1967.

Ragosta, Marjorie, Paul Holland, and Dean T. Jamison. *Computer-Assisted Instruction and Compensatory Education: The ETS/LAUSD Study.* Princeton, N.J.: Educational Testing Service, 1981.

Riel, M. "The Computer Chronicles Newswire: A Functional Learning Environment for Acquiring Literacy Skills." *Journal of Educational Computing Research* 1 (1985): 317–37.

Rosengren, Karl S., Dana Gross, Anne F. Abrams, and Marion Perlmutter. "An Observational Study of Preschool Children's Computing Activity." Paper presented at the "Perspectives on the Young Child and the Computer" conference, University of Texas at Austin, September 1985.

Sandall, Susan R., Mary E. McLean, and Barbara J. Smith, eds. *DEC Recommended Practices in Early Intervention/Early Childhood Special Education.* Denver, Colo.: Division for Early Childhood of the Council for Exceptional Children, 2000.

Sarama, Julie. "Redesigning Logo: The Turtle Metaphor in Mathematics Education." Ph.D. diss., State University of New York at Buffalo, 1995.

Sarama, Julie, and Douglas H. Clements. "Learning and Teaching with Computers in Early Childhood Education." In *Contemporary Perspectives in Early Childhood Education,* edited by Olivia N. Saracho and Bernard Spodek, pp. 171–219. Greenwich, Conn.: Information Age Publishing, 2002.

Sarama, Julie, Douglas H. Clements, and Elaine Bruno Vukelic. "The Role of a Computer Manipulative in Fostering Specific Psychological/Mathematical Processes." In *Proceedings of the Eighteenth Annual Meeting of the North America Chapter of the International Group for the Psychology of Mathematics Education,* edited by Elizabeth Jakubowski, Dierdre Watkins, and Harry Biske, pp. 567–72. Columbus, Ohio: ERIC Clearinghouse for Science, Mathematics, and Environmental Education, 1996.

Schery, Teris K., and Lisa C. O'Connor. "Language Intervention: Computer Training for Young Children with Special Needs." *British Journal of Educational Technology* 28 (1997): 271–79.

Shade, Daniel D. "Computers and Young Children: Software Types, Social Contexts, Gender, Age, and Emotional Responses." *Journal of Computing in Childhood Education* 5, no. 2 (1994): 177–209.

Shade, Daniel D., Robert E. Nida, Judith M. Lipinski, and J. Allen Watson. "Microcomputers and Preschoolers: Working Together in a Classroom Setting." *Computers in the Schools* 3 (1986): 53–61.

Shields, Margie K., and Richard E. Behrmann. "Children and Computer Technology: Analysis and Recommendations." *The Future of Children* 10, no. 2 (2000): 4–30.

Shrock, Sharon A., Margaret Matthias, Juliana Anastasoff, Cyndi Vensel, and Sharon Shaw. "Examining the Effects of the Microcomputer on a Real World Class: A Naturalistic Study." Paper presented at the meeting of the Association for Educational Communications and Technology, Anaheim, Calif., January 1985. (ERIC Document Reproduction Service No. ED256335)

Silvern, S. B., T. A. Countermine, and Peter A. Williamson. "Young Children's Interaction with a Microcomputer." *Early Child Development and Care* 32 (1988): 23–35.

Steffe, Leslie P., and Heide G. Wiegel. "Cognitive Play and Mathematical Learning in Computer Microworlds." *Journal of Research in Childhood Education* 8, no. 2 (1994): 117–31.

Stone, Theodore T., III. "The Academic Impact of Classroom Computer Usage upon Middle-Class Primary Grade Level Elementary School Children." Ed.D. diss., Widener University, 1996. *Dissertation Abstracts International* 57-06 (1996): 2450.

Subrahmanyam, Kaveri, Robert E. Kraut, Patricia M. Greenfield, and Elisheva F. Gross. "The Impact of Home Computer Use on Children's Activities and Development." *The Future of Children* 10 (2000): 123–44.

Turkle, Sherry. "Seeing through Computers: Education in a Culture of Simulation." *American Prospect* 31 (1997): 76–82.

Watson, J. Allen, and Vicki M. Brinkley. "Space and Premathematic Strategies Young Children Adopt in Initial Logo Problem Solving." *Journal of Computing in Childhood Education* 2 (1990/91): 17–29.

Watson, J. Allen, Garrett Lange, and Vicki M. Brinkley. "Logo Mastery and Spatial Problem-Solving by Young Children: Effects of Logo Language Training, Route-Strategy Training, and Learning Styles on Immediate Learning and Transfer." *Journal of Educational Computing Research* 8 (1992): 521–40.

Wright, June L. "Listen to the Children: Observing Young Children's Discoveries with the Microcomputer." In *Young Children: Active Learners in a Technological Age,* edited by June L. Wright and Daniel D. Shade, pp. 3–17. Washington, D.C.: National Association for the Education of Young Children, 1994.

Xin, Joy F. "Computer-Assisted Cooperative Learning in Integrated Classrooms for Students with and without Disabilities." *Information Technology in Childhood Education Annual* 1, no. 1 (1999): 61–78.

Yelland, Nicola. "A Case Study of Six Children Learning with Logo." *Gender and Education* 6 (1994): 19–33.

Yelland, Nicola J. "Making Sense of Gender Issues in Mathematics and Technology." In *Gender in Early Childhood,* edited by Nicola J. Yelland, pp. 249–73. London: Routledge, 1998.

Yost, Nancy Jill Mckee. "Computers, Kids, and Crayons: A Comparative Study of One Kindergarten's Emergent Literacy Behaviors." Ph.D. diss., Pennsylvania State University, 1998. *Dissertation Abstracts International* 59-08 (1998): 2847.

5

Comparing Distributions and Growing Samples by Hand and with a Computer Tool

Arthur Bakker
Ann Frederickson

A LITANY of research in statistics education shows that students often learn statistics as a set of techniques that they are not able to apply sensibly. For example, it is well documented that having learned to calculate the mean, students often do not understand that they can use the mean as a group descriptor when comparing two data sets (Mokros and Russell 1995). Although the measures of center are not at all easy to understand and use (Zawojewski and Shaughnessy 2000) and students often miss the conceptual underpinnings of the mean (Konold and Pollatsek 2002), many students calculate the mean whenever a problem sounds statistical, even if it is not very useful.

Looking for Different Ways to Teach Statistics

We can compare this situation to the proverbial tip of the iceberg. It is the substance beneath the surface that makes the iceberg float. In this metaphor, mean, median, and mode are the visible tip of the iceberg. What is beneath the visible surface is the knowledge and skills that students really need to understand and sensibly use these measures of center. In looking for statistical concepts that make sense to students, many teachers and researchers have been focusing on "key concepts." At the middle school level we can think of variation, sampling, data, distribution, and center as being key concepts. In our view, students should learn to reason about these key concepts in a coherent way, starting with their own informal statistical notions and graphs and

We thank Truus Dekkers, Katie Makar, Meg Meyer, and Monica Wijers for their helpful comments. We thank Eena Khalil and Cliff Konold for letting us use their fish data cards. The research was funded by the National Science Foundation under grant number ESI-9818946. The opinions expressed in this paper do not necessarily reflect the views of the Foundation.

Editor's note: The CD accompanying this yearbook contains a Tinkerplots file that is relevant to this article. For more information on Tinkerplots, contact Key Curriculum Press at www.keypress.com/. Tinkerplots screenshots in this article are reproduced, with permission, from TinkerPlots™ Dynamic Data Exploration, Key Curriculum Press, 1150 65th Street, Emeryville, CA 94608, 1-800-995-MATH, www.keypress.com.

working toward more conventional ones. Central among the critical skills students need to acquire are comparing distributions and viewing stable features in the variation of samples. The question we answer in this article is how such coherent reasoning about statistical key concepts can be promoted.

In this paper we discuss instructional activities that can engage students in reasoning about statistical key concepts: comparing distributions and growing samples. The core idea behind growing a sample is predicting what shape a graph will have and how a distribution emerges when more data are added (Konold and Pollatsek 2002). We recount how the activity of growing samples by hand was carried out in two sixth-grade classes and how this formed the basis for activities with a computer tool for data analysis.

We start with a discussion of key concepts in middle school statistics. We then examine students' ideas as they engage in an instructional activity on growing samples. Finally, we reflect on how technology can be used to support students in analyzing data.

Key Concepts at the Middle School Level

In this section we discuss the key concepts of variation, sampling, data, distribution, and center.

Variation is at the core of all statistical investigation. For example, consider the context of how long batteries last. If students do not expect any variation in the life span of batteries, they would not have any reason to take a sample or look at a distribution. Variation as a topic did not receive much attention in statistics education research until about 1997. The reason for this neglect was probably because the focus was usually on the measures of center. There are different types of variation, such as variation around a mean, variation in frequency, variation around the smooth curve of the normal distribution, variation within and among distributions, variation among samples, and covariation.

Sampling. To describe or predict a particular variable phenomenon, we need data, and data are mostly created by taking a sample.

Data. Understanding the key concept of data includes insight into why data are needed and how they are created. An understanding of the concept of data relies on knowledge about measurement.

Distribution. When people search for patterns and trends, they use graphs to find relations and characteristics that are not visible from a table. The problem with finding such relations and characteristics, however, is that many students tend to perceive data as a batch of individual cases rather than as a whole that has characteristics that are not visible in any of the individual cases. Hancock, Kaput, and Goldsmith (1992) note that students need to construct such an aggregate mentally before they can perceive a data set as a whole. To move from a case-oriented view to an aggregate view of data, students need to develop a conceptual structure with which they can conceive of data sets as aggregates. The concept of distribution is such a structure. In the words of Petrosino, Lehrer, and Schauble (2003, p. 132): "Distribution could afford an organizing conceptual structure for thinking about variability located within a more general context of data modeling." It is possible and desirable to address the issue of how data are distributed from an informal situational level onwards by focusing on shape. In the research of Cobb, McClain, and Gravemeijer (2003), students came to reason about hills as indicators of majorities in data sets. Bakker and Gravemeijer (2004) report on how students reasoned about data on students' weight and how a "bump" (the distribution) changed if the sample size grew and how it would shift if older students' weights were measured. In this way, students extended their case-oriented view with an aggregate view on data.

Center. Traditionally, measures of center such as mean and median are taught before students have developed a notion of center. Zawojewski and Shaughnessy (2000) state that the reason students find mean and median difficult is because they have not had sufficient opportunities to make connections between center and spread; that is, they have not made the link between measures of central tendency and the distribution of the data sets. It is possible to choose sensibly between mean and median only if one takes the context and the distribution of the data into account. It is therefore important to give students opportunities to learn ways to describe how data are distributed, even before teaching formal measures of center.

How can students develop an understanding of these key concepts? We have designed several activities to engage students in statistical investigations that would involve these concepts at an informal level. In this article, we show how sixth-grade students reasoned about variation and distribution aspects when comparing distributions and when thinking about growing samples.

Background of the Classes

For the remainder of this article, we report on two sixth-grade classes from a public middle school in a suburban area in the Midwest. In one class, twelve of the twenty-two students had at least one label such as "English as a second language" (7), "extra reading help" (5), or "at risk" (3). In the other class, only one student had English as a second language. They had all studied mean, median, and mode in grade 4 or 5.

In the first lesson, students collected data about students' foot length and had made a so-called graph-feet-ee on the wall in the corridor (fig. 5.1). In the first discussion, students tended to look only at the mode or the highest and lowest values in the graph. Students in the first class tried to calculate

Fig. 5.1. Discussion of the "graph-feet-ee"

the mean but did not succeed. Some of them were able to find the median foot in the graph, but they could not explain whether the result of this procedure was Harry, his foot, or the value 23 cm. This was additional evidence that we should not focus on teaching measures of center as formal statistical techniques but rather step back and work on developing an appropriate conceptual basis. What useful informal ideas of center do these students have that can be developed into more-formal statistical notions? We decided to let students develop a language in which they could talk about aggregate aspects of shapes of distributions. In previous research, Bakker and Gravemeijer (2004) noticed that seventh and eighth graders tended to conceive of data sets as groups of low, "average," and high values. Students used informal terms such as *clumps* and *majorities* to indicate these "average" groups, which often functioned as an indicator of center. For students, these "clumps" seem to be more intuitive and meaningful summaries of data sets than *mean* or *median* (Konold et al. 2002).

From the second lesson onward, students used a computer tool for data analysis called Tinkerplots (Konold and Miller 2004), which is especially designed for middle school students. In this software, a data set is first shown as an unorganized set of data cards (fig. 5.2). Using basic operations such as separate, stack, and order, students can organize the data set into plots such as a bar graph or a dot plot (which is also called a number-line plot). Once the data

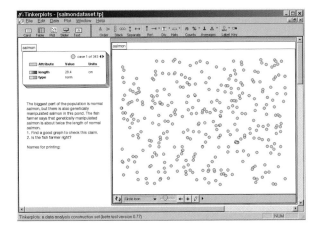

Fig. 5.2. The opening screen of Tinkerplots showing a large data set

cards have been organized along an axis, it is possible to make more-advanced plots such as a histogram or a box plot. With this bottom-up tool for constructing graphs, students can explore data sets with multiple representations—both unconventional and conventional—in ways that are not feasible by hand.

Students were enthusiastic about the software. We heard a lot of "wow, cool, awesome, that's me!" although they did not always understand what the software was doing. To stimulate reflection, we asked students to write their observations on index cards and share their discoveries in class discussions. The software was used in about one-third of the class periods. Before these class periods, students compared distributions of different situations, such as the braking distances of cars, the life spans of batteries, and the heights of fathers and sons. They gradually developed a language with which they could describe variation and center issues. For instance, they typically used terms such as *spread out* or *bunched up* to characterize the variation within a distribution, and after about twelve class periods, they used the notions of "majority" or "middle clump" to summarize the centers of data sets. Students either pointed to the clump's position or mentioned the range of the clump. Although the clumps were not well defined, reasoning with them seemed to help students focus on aggregates rather than on individual cases. In addition, these "clumps" appeared to function better as an intuitive notion of center than the formal measures of center. This discussion provides some background to the activity of growing samples, which lasted one class period of forty-two minutes.

Story of a Fish Farmer

According to the story used for this activity, a fish farmer grows genetically engineered (GE) fish. He claims that these fish grow bigger than normal fish. One year after releasing a bunch of normal fingerlings and a smaller number of GE fingerlings into a pond, students are allowed to catch some fish to check his claim.

Each student simulated "catching" about four fish from the pond by drawing cards from a box (fig. 5.3). Normal fish were represented by yellow data cards, whereas the GE fish data cards had a slightly lighter color. The length of each fish was shown on the data card. Each student had an activity sheet with axes on which they could plot the data. With the students sitting in groups of four, we expected that they would see the variation between the samples, both in length and the type of fish. One student was indeed disappointed that he had caught only normal fish, whereas other students had caught both kinds of fish.

Fig. 5.3. Drawing a sample

Students took turns going up and plotting their own data on the whiteboard while the teacher read their fish lengths aloud (fig. 5.4). At the same time, every student plotted the data of all students on their own activity sheet, thereby allowing every student the opportunity to experience how the sample grew. We noted that students were very concerned with accuracy. It was apparent that they tended to focus on individual data values and not on general features or shapes, which was our purpose. After a couple of turns, students collectively had a closer look at the intermediate result (fig. 5.5).

Fig. 5.4. Plotting the data on the whiteboard

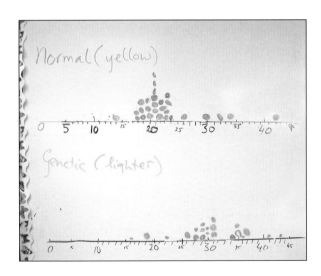

Fig. 5.5. Sample of the first class

Comparing Distributions

After taking the variation in sample size and the variation within the two distributions into account, an expert might use the means of the normal and the GE fish data sets to compare the sizes of the two types of fish. What do students do if we do not push them to use means? In this section we offer some examples of how students reasoned about the two distributions and how they dealt with different types of variation.

With reference to figure 5.5, we first asked for observations, which yielded a variety of answers. Several students indicated where they saw the clumps in the two distributions, for example, "The clump [of the normal fish] is in between 18 and 23 [cm]." One boy said that the normal fish were more spread out than the GE fish, but he also noticed that there were fewer GE fish. One girl used the clump notion to compare the two types of fish while taking the sample size into account.

> *Linda:* Right here [*pointing to the normal fish data*] there is a bigger clump than this one [GE], this one, but the numbers are less. This one [normal], the clump is more spread out but higher.
>
> *Researcher:* You said a *lot* of things. So one was ….
>
> *Linda:* This [normal] clump is bigger [more values, i.e., more fish] but has a smaller number [lower values, i.e., shorter lengths]. And this one [GE] is smaller [fewer values, i.e., fewer fish], but it has a higher number [higher values, i.e., longer lengths].

The examples show how students were developing a language with which they could describe different types of variation: between the distributions (GE fish are longer), and between the sample size (the GE fish form a smaller sample, which also causes the lower height of the clump).

Growing Samples

In this section we focus on students' predictions about what would happen if they made the sample bigger or if "the sample grew." The sample in the second class looked a little different (fig. 5.6).

Students made several conjectures about what would happen with larger samples. Barbara expected to get a clump between 20 cm and 35 cm. Anissa said, "The top is a little more spread out, but that may be because there are more." One student thought that the clump would stack up; another thought

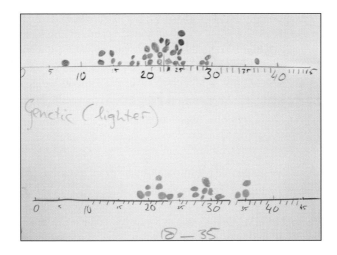

Fig. 5.6. The sample discussed in the second class

that the clump would spread out. Sheila said, "I think it will stay in the same area." This is one of the insights we aimed at: coming to see stable features in variable processes. In this example, Sheila expected stability in the area where new data values would appear.

One of the important goals of the instruction was that students would come to conceive of aggregate features of data sets as opposed to just individual cases. One of the ways to stimulate an aggregate view was to focus on general shapes of data sets, which stabilize if the sample size becomes big enough. We therefore asked students to draw a sketch of what they would expect. Some (but not all) students had reasonable intuitions about the shape of the distribution. For example, Norman compared the shape with that of fathers' heights (a normal distribution), an example he knew from an earlier investigation. When asked why, he responded as follows.

Norman:	'Cause that is kind of what it is forming now, and there couldn't be much different fish. If that is a sample, a good sample, then it'll probably just get bigger but stay in the same shape....
Researcher:	Norman, could you stand up and explain that kind of shape?
Norman:	Oh, that's how I kind of thought it looked. And I just tried to make it bigger. [This could mean that he already saw that shape in the small sample and just made it bigger.] I could

have gotten more detail, but this is kind of a rough sketch [thus reducing variation in the height of the shape].

Researcher: ...What would we get as a shape? What kind of shape do you see?

Kerry: A kind of bumpy slope. Gradually climbing up.

Researcher: And what if we had the whole population, all the fish. What shape would it have? Would it still be bumpy, or ...?

Kerry: Yeah. It'd be a bit more smooth, I think.

After discussing the shape, we returned to the original question of whether the GE fish would grow bigger than the normal fish.

Anissa: I would say that there is more of a variety in the normal fish, but I would actually say that the genetic fish do grow taller, because more of them are closer to the, how do you say, the right side.... Up there is four really close to it. This one and these three are pretty close to the high end. So I think he [the fish farmer] is right.

The interesting thing about her answer is that she referred to the GE fish in general, but in her explanation she seemed to adopt a case-oriented view. We went on to ask:

Researcher: How could we find out how big the difference is? If there is a difference....

Norman: What we have to do is go on to Tinkerplots, put all the information on Tinkerplots, then make a graph and find the mean value, and use a reference line, and find out, uhm, how far both mean values are apart, if they are apart, and then it'll probably, however far they are apart, that's, uhm, how much bigger they are.

From such episodes we concluded that their experience with the software helped some students to develop language and thinking tools with which they could solve statistical problems. In the next period, students were to analyze another fish data set, but now with Tinkerplots software.

Follow-up with a Computer Tool

For the next activity, we used a different data set. The question was whether the fish farmer was right in claiming that GE fish were about twice as big as the normal fish. If students open the data set, they see only a stack of data cards (upper left corner in fig. 5.2) and an unorganized set of data icons (the plot at the right). Students can organize these data icons by separating, stacking, or ordering these icons in relation to different variables, in this example by the type of fish and length. The possible plots include dot plots, bar graphs, histograms, and box plots as well as many unconventional plots. In previous lessons, students had mainly used dot plots.

Students spent most of their time in the computer lab finding a plot they found useful. Class discussions were conducted to evoke reflection on these plots. One plot that many students made was a precursor to the histogram: dots were separated into intervals and stacked (fig. 5.7). Students referred to this plot as the "Rick plot," since Rick had demonstrated it during a class discussion.

Some students also fused the dots into bars, a procedure that yields histograms (fig. 5.8). Another plot that many students made was a value-bar graph (fig. 5.9). In this plot, Karen chose to represent the data as value bars, whose relative lengths correspond to their values. She then ordered the value bars by their value from short to long. With the reference lines set at the maximum values, she argued that the GE fish were larger than the normal fish (fig. 5.9). "I think the fish farmer should be more specific. If you compare the smallest normal fish and the largest genetic, then it would be correct but [the

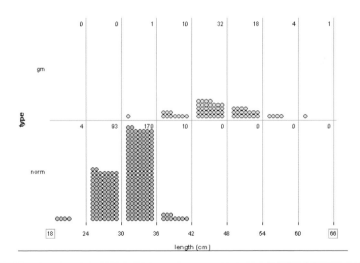

Fig. 5.7. The "Rick plot." Rick had separated the dots into bins of width 6 cm and had stacked them vertically. This is a precursor to the histogram. Then the "n" button gave the number of dots in each bin.

other way round] then he was wrong." We consider this a case-oriented view.

For some students, the center clump became an aggregate reasoning tool. Kerry, for instance, compared the two clumps "28–35 and 42–49, which is not twice as much." Moreover, students started to react more to one another than just to the teacher—which shows their engagement.

> *Karen:* If you say "clump," do you mean where most of the fish are?
> Or do … [*interrupted*].
>
> *Kerry:* Yeah, like where most of the fish are normal, their length is
> in between the range 28–35, where most of them are.

Tom's solution came closer to a conventional approach. Using the soft-

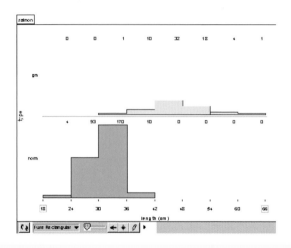

Fig. 5.8. Histograms. When "Fuse Rectangular" is selected, the dots fuse into bars, which gives a histogram.

ware, he separated the fish types vertically, ordered the lengths horizontally, stacked the dots, and used the mean button and reference lines to compare the types of fish (fig. 5.10).

His explanation gives an impression of how some students learned to compare distributions:

> *Tom:* I clicked the mean value and the reference line, because it
> shows kind of where the clump is. And that helps me because
> it is easier for me to see where most of them are. And this one
> [the normal fish] there is a lot more, there is 292 [using the

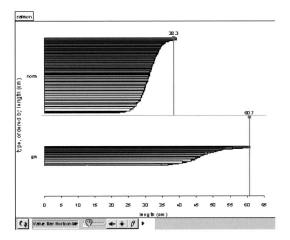

Fig. 5.9. The fish are lined up according to their length. Karen used the maximum values to compare the groups (not the means).

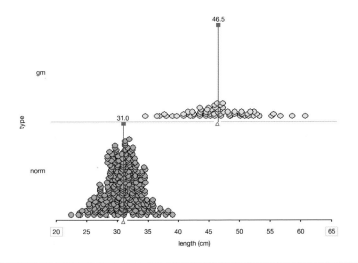

Fig. 5.10. Tom's solution. The blue triangles give the position of the mean values, and the movable vertical red lines are called reference lines.

count option]. And this one, there is 67, so there is about a fourth, a little under a fourth. This one [GE] is a lot more gradual, it is spread out, but they grow a lot bigger and this

one [normal] is very steep [points with the mouse along the
slope] and then is really steep, going down. And you can see
that it is not really twice. These ones aren't twice the size of
these ones. It is more like one and a half times.

It is interesting that Tom first used the means to indicate where the clumps
were. This was what we were after: first let students develop a notion of center (for instance, with clumps) and then let them measure it more formally
with mean or median. In that way, the mean is used as a group descriptor,
which could be a sign of an aggregate view on data. After Tom had explained
his solution, he nevertheless felt the need to compare the smallest and highest values. This could indicate a case-oriented view on the data, but it also
might indicate a sense that looking at the range of the values is important.

Using the Software

Using this data analysis software turned out to be rewarding for students.
First of all, they were very motivated to use it. In addition, it offered the possibility to explore multiple representations in search of a meaningful and convincing plot with which they could answer the question at issue. In this way
they were able to analyze data sets that would be much too large for analyzing by hand. Additionally, they could use it to calculate quickly means, medians, and so on.

Tinkerplots has been designed as a construction tool for middle school students with which they could make their own plots, but how do students
describe it? To the visiting superintendent, students explained what
Tinkerplots was in the following way:

Kerry: It is sort of like a graphing sort of system that you can use
to figure out that you can make graphs out of and actually it
is pretty fun, too. It is sort of like, … it's got the same sort
of things as any other sort of graphing thing but you actually make the graph yourself instead of entering all the numbers in it and the computer does it for you. Like there is a
bunch of dots, a bunch of little dots usually, like turn in the
menu and make a rectangle out of it. It is really fun.

Carl: You can make graphs like bar graphs and line graphs [dot
plots], and then you would take the graph that you can
answer your [question with] and you can have different buttons and you can push, like Kerry said, mean and median
and then you can find out a lot of stuff about your data.

Although the students were enthusiastic about the software, it was not always easy for the teacher to support students' learning of data analysis. We noticed that focused reflection was extremely important—whether in small groups or in whole-class discussions. Why is a particular plot helpful in answering a question? Are there better ways? Do you understand what Kerry is saying?

We have been wondering how well students would have worked with the software without drawing samples and plotting the data values by hand. We assume that growing samples by hand forms the basis for exploring data sets with a computer tool. The published version of Tinkerplots has even more capabilities than the prototype we used. For instance, students can grow samples with a slider (fig. 5.11). Thus they can dynamically interact with a data set to explore how the mean or the shape of the distribution changes with the sample size. It is also possible to take different samples of the same size (resampling) and compare the changes in shape. One of the things that teachers can discuss with their students is the stability of the mean or shape across sample sizes.

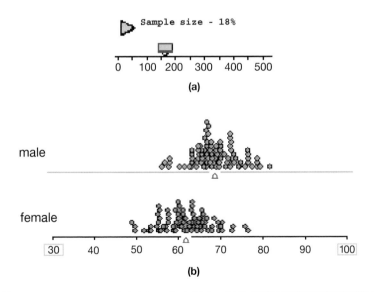

Fig. 5.11. Using sliders (a) to grow samples and watch the stability of the mean, median, or shape (b)

Conclusions

Statistics education may seem to be about mean, median, mode, range, quartiles, bar graphs, histograms, and so on. Throughout this article we have argued that such a set of statistical notions and graphs is only the tip of the statistics iceberg. Students need to spend considerable time in developing a bigger picture of statistics. To understand mean and median requires a notion of the center of a distribution. Before students can sensibly use formal measures of center in comparing distribution, they first need to understand the many types of variation they encounter: variation in and among distributions, and variation in sample size and frequency. They even need to have a sense of the shape of the data, because not all data sets are suitable to compare with means or medians. If we teach such formal measures too early, students may be inclined to use such measures without looking at variation and distribution, which could lead to superficial and incorrect conclusions. In our view, a narrow focus on mean, median, and mode is an example of premature formalization.

Why do almost all school textbooks follow the same sequence and introduce mean, median, and mode together, providing students with graphical tools like histograms and box plots long before students have the conceptual understanding to use such tools sensibly? One important reason could be that mean, median, mode, and graphs seem so easy to teach and to assess. The approach taken in this article is challenging for the teacher because students' notions stay informal or preformal for quite a while. Their learning with such an approach might be harder to assess than if they just learn to calculate average values and to draw histograms.

A computer tool offers the opportunity to explore large data sets in a way that is not feasible by hand. Interacting with plots such as dot plots may give students a feeling for the shape of a data set, which is not visible from a table or from a mean. However, we recommend plotting and analyzing data by hand as well in order to provide a meaningful context for plotting and analyzing data with a computer tool.

In this article we mapped out a few conceptual underpinnings of center and variation that can help students develop a bigger picture of statistics. Comparing distributions and growing samples in the way described in this article, both with and without a computer tool, appear to contribute to such a bigger conceptual picture of statistics and data analysis.

REFERENCES

Bakker, Arthur, and Koeno P. E. Gravemeijer. "Learning to Reason about Distribution." In *The Challenge of Developing Statistical Literacy, Reasoning, and Thinking,* edited by Dani Ben-Zvi and Joan Garfield, pp. 147–67. Dordrecht, Netherlands: Kluwer Academic Publishers, 2004.

Cobb, Paul, Kay McClain, and Koeno P. E. Gravemeijer. "Learning about Statistical Covariation." *Cognition and Instruction* 21, no. 1 (2003): 1–78.

Hancock, Chris, James J. Kaput, and Lynn T. Goldsmith. "Authentic Enquiry with Data: Critical Barriers to Classroom Implementation." *Educational Psychologist* 27, no. 3 (1992): 337–64.

Konold, Clifford, and Craig Miller. Tinkerplots. Data analysis software for middle school curricula. Emeryville, Calif.: Key Curriculum Press, 2004.

Konold, Clifford, and Alexander Pollatsek. "Data Analysis as the Search for Signals in Noisy Processes." *Journal for Research in Mathematics Education* 33 (July 2002): 259–89.

Konold, Clifford, Amy Robinson, Khalimahtul Khalil, Alexander Pollatsek, Arnold D. Well, Rachel Wing, and Susanne Mayr. "Students' Use of Modal Clumps to Summarize Data." In *Proceedings of the International Conference on Teaching Statistics [CD-Rom], Cape Town,* edited by Brian Phillips. Voorburg, Netherlands: International Statistics Institute, 2002.

Mokros, Jan, and Susan Jo Russell. "Children's Concepts of Average and Representativeness." *Journal for Research in Mathematics Education* 26 (January 1995): 20–39.

Petrosino, Anthony J., Richard Lehrer, and Leona Schauble. "Structuring Error and Experimental Variation as Distribution in the Fourth Grade." *Mathematical Thinking and Learning* 5, no. 2 & 3 (2003): 131–56.

Zawojewski, Judith S., and J. Michael Shaughnessy. "Mean and Median: Are They Really So Easy?" *Mathematics Teaching in the Middle School* 5 (March 2000): 436–40.

Interactive Geometry Software and Mechanical Linkages: Scaffolding Students' Deductive Reasoning

Jill Vincent

ALTHOUGH Euclidean geometry and geometric proof once occupied a central place in mathematics education, research indicates that many students now fail to understand the purpose of mathematical proof, are unable to construct proofs, and readily base their conviction on empirical evidence or the authority of a textbook or teacher. In the light of such evidence, what is the rationale for including proof in school mathematics, and how can proof be made more accessible to students?

In the words of one grade 8 student, "proof is the concrete base of a house built of mathematics." Too often, though, school mathematics focuses more on the product than the process and fails to convey how mathematics has evolved as a logical system founded on proof. According to the Australian Education Council (1991, p. 14),

> Mathematical discoveries, conjectures, generalisations, counter-examples, refutations and proofs are all part of what it means to do mathematics. School mathematics should show the intuitive and creative nature of the process, and also the false starts and blind alleys, the erroneous conceptions and errors of reasoning which tend to be a part of mathematics.

Proof helps students to see mathematics as the result of human endeavor and as a logically constructed discipline rather than as a series of unrelated esoteric theorems and rules. Proof establishes connections, and it can deepen

I would like to thank Helen Chick of the University of Melbourne and Barry McCrae of the Australian Council for Educational Research for their guidance, support, and much-valued constructive criticism throughout all aspects of the research. I also wish to express my gratitude to the Year 8 students for their enthusiastic approach to the research tasks.

Editor's note: The CD accompanying this yearbook contains eleven Cabri files that are relevant to this article. For more information on Cabri, contact Cabri at www-cabri.imag.fr/cabri/index-e.html or at education.ti.com/us/product/software/cabri/features/features.html.

the understanding of mathematical concepts. Goldenberg, Cuoco, and Mark (1998, p. 42) assert that

> proof is not merely to support conviction, nor to respond to a distrustful nature of self-doubt, nor to be done as part of an obsessive ritual. Proof serves to provide explanation.

The promotion of proof as a process through which mathematical knowledge and understanding have been constructed will not necessarily motivate students, though, unless they believe that they are participating in meaningful mathematical discovery. As Hanna and Jahnke (1993, p. 433) note,

> the challenge of actual mathematical enquiry cannot be entirely reproduced. All the parties to the classroom interaction, teachers and students, know that they are dealing with theorems that have already been proven by others.

Today's challenge, then, is to design tasks where students experience a genuine cognitive need for conviction and where proving offers them the satisfaction of understanding *why* their conjectures are true. Reflecting this relationship between proof and understanding, mathematics curricula in many countries (see, for example, National Council of Teachers of Mathematics 2000) currently are emphasizing the need for students to justify and explain their reasoning. In geometry, however, concern has been expressed that interactive geometry software, such as Cabri Geometry II and The Geometer's Sketchpad, may be contributing to a data-gathering approach, where empirical evidence is becoming a substitute for proof. Noss and Hoyles (1996) assert that in the United Kingdom, for example, traditional geometry exercises are being adapted for the computer, and geometry is being reduced to spotting patterns in data generated by dragging and to measuring screen drawings, with little or no emphasis on theoretical geometry: "school mathematics is poised to incorporate powerful dynamic geometry tools in order merely to spot patterns and generate cases" (p. 235). Hölzl (2001, pp. 68–69) suggests that the problem lies with the way interactive geometry software is used rather than with the software itself:

> The often mentioned fear that the computer hinders the development of an already problematic need for proof is too sweeping. It is the context in which the computer is a part of the teaching and learning arrangement that strongly influences the ways in which the need for proof does—or does not—arise.

A Teaching Experiment

The Context

In the quest for a motivating context in which to introduce grade 8 (twelve-to-thirteen-year-old) students to geometric proof, my attention was drawn to mechanical linkages, or systems of hinged rods (see Vincent and McCrae 2001; Vincent, Chick, and McCrae 2002). Linkages are found in many common household items, for example, folding umbrellas, expanding toolboxes, and car jacks, as well as in old drawing instruments referred to as pantographs. Many are based on simple geometry such as similar figures, isosceles triangles, parallelograms, or kites, and they can be constructed readily from plastic geometry strips (geostrips) and paper fasteners. Interactive geometry software is particularly suitable for modeling mechanical linkages, since the action of a linkage can be simulated by means of the drag and animation options.

The Participants

The participants in the teaching experiment were twenty-nine grade 8 girls who, on the basis of their grade 7 mathematics performance, represented the upper 25 percent of their grade level at the school. During the previous year the students had followed the standard geometry component of the mathematics curriculum at the school, which included side and angle properties of triangles and quadrilaterals, and polygon angles. All students had their own notebook computers with interactive geometry software, but they had not used the software in grade 7. As a member of the teaching staff of the school, I was both the researcher and the regular mathematics teacher for the grade 8 class.

At the start of grade 8, geometry that would be required in the conjecturing-proving tasks was taught or revised: properties of triangles and quadrilaterals, angles associated with parallel lines cut by a transversal, the Pythagorean theorem, and similar and congruent triangles. During these lessons, the students engaged in exploratory activities with interactive geometry, but the emphasis was deliberately on empirical data and identifying properties, with no reference to why these properties were true. This was important so that conjecturing and argumentation in geometry, and the concept of proof, would be new experiences for the students when they came to the lessons associated with the teaching experiment. The students had limited experience with the software's Tabulation facility, and they were aware of the Trace option, which allows the locus of a selected point to be traced when a screen construction is dragged, but they had not used this tool in any specific tasks.

Prior to the commencement of the conjecturing-proving lessons, a 48-item van Hiele test (Lawrie 1997) covering six geometric concepts—squares, right-angled triangles, isosceles triangles, parallel lines, congruency, and similarity—was administered to assess the students' levels of understanding of properties and relationships. Although the students in this grade 8 class were regarded as the top 25 percent of grade 8 students in mathematics in their school, the van Hiele test showed that they were in no way exceptional as a group with respect to their geometric understanding. Twelve of the twenty-nine students were assessed at van Hiele Level 1 for at least one of the six concepts, and only five students had reached Level 3 for four or more concepts.

The Lesson Sequence

The research took place over eighteen 50-minute lessons in a six-week period, during which the students were introduced to the concept of geometric proof and participated in a number of conjecturing-proving tasks. Seven of these tasks were based on mechanical linkages, and in each task the students worked with a physical model of the linkage, referring to worksheet diagrams to construct their own linkages from plastic strips and paper fasteners. In instances where real mechanical linkages were available, for example, the car jack, the students also worked with these. The students seemed to enjoy working with the geostrip linkages as much as with real linkages, so it was in no way a disadvantage that they did not have access to real mechanical linkages in all situations. The students were also provided with a teacher-prepared interactive geometry simulation of each linkage to use on their notebook computers. In addition to the mechanical linkage tasks, the students completed a paper-and-pencil triangle-midpoints proof and two interactive geometry exploratory conjecturing-proving tasks—quadrilateral midpoints and circle angles.

Setting the Scene for Proof

Following the Industrial Revolution, a number of mathematicians became involved in designing mechanical linkages that converted circular motion into straight-line motion. Some of these produced genuine linear motion, whereas others produced only approximate linear motion over part of their movement. One of these, Tchebycheff's linkage (see fig. 6.1; Bolt 1991), was used to introduce the students to conjecturing and to establish the need for proof. The linkage consists of three rigid hinged bars, *AC*, *BD*, and *CD*, with lengths 5, 5, and 2 units respectively. Points *A* and *B* are fixed, with the distance *AB*

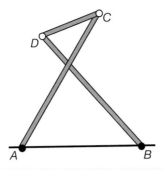

Fig. 6.1. Tchebycheff's linkage

equal to 4 units. The students constructed the linkage from geostrips, using paper fasteners to attach points A and B to a piece of firm paper. They could then rotate C and D and trace the paths of different points on the linkage by placing a pencil through holes in the geostrips. From their observations with the geostrip linkage, the students conjectured that the midpoint of CD moved in a straight line.

When the students were given access to a computer model of the linkage (see fig. 6.2), their realization that the path of P (the midpoint of $\overline{CD}$) was not linear and their astonishment at seeing how little the path actually deviated from a straight line were sufficient to convince them that visual evidence

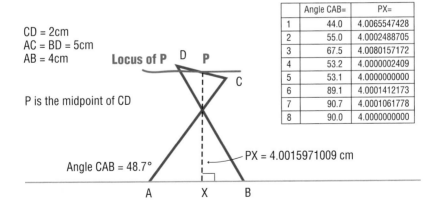

Fig. 6.2. Interactive geometry model of Tchebycheff's linkage

could not be trusted. Furthermore, the tabulated values for the distance *PX* showed that empirical evidence could not always be relied on either—the measurement of the distance to only one or two decimal places could have led to the false conclusion that the distance *PX* was invariant. One could argue that in principle this approach was little different from traditional textbook use of optical illusions to demonstrate a need for proof. What seemed to be significant, however, was that the students themselves had been actively involved in the generation of the false conjecture.

Modeling a Proof Construction

The concept of proof as a convincing argument was then introduced, noting that an argument based on empirical evidence would not convince everyone. A proof—that the sum of the angles of any triangle is 180°—was demonstrated (see fig. 6.3), and this proof became the model for the students' proof writing. The emphasis was on justifying deduced statements instead of presenting a rigorous formal proof.

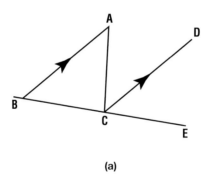

(a)

Given: BCE is a straight line, DC // AB

Prove: $\angle ABC + \angle BAC + \angle ACB = 180°$

Proof: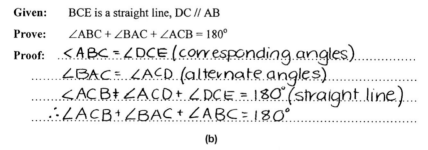

$\angle ABC = \angle DCE$ (corresponding angles)

$\angle BAC = \angle ACD$ (alternate angles)

$\angle ACB + \angle ACD + \angle DCE = 180°$ (straight line)

$\therefore \angle ACB + \angle BAC + \angle ABC = 180°$

(b)

Fig. 6.3. Modeled proof for the sum of the angles of any triangle

Conjecturing-Proving Tasks

The students worked in pairs on each conjecturing-proving task. During the argumentations associated with each task, the teacher's intervention fulfilled several important roles: clarifying the content and probing the meaning of the students' statements, answering queries, correcting false statements, and redirecting the students' thinking if they had reached an impasse. The most important aspect of teacher mediation, however, was to ensure that the students' arguments were based on sound mathematical logic. As the students' regular mathematics teacher, I also considered it important that a lesson should not end without some sense of achievement and at least partial success in the conjecturing-proving process, and this belief brought about some of the interventions.

To illustrate the students' progress during the conjecturing-proving tasks, the focus will be on four linkage tasks completed by one pair of students—Anna and Kate—although references will be made to other students.

Car Jack

The first conjecturing-proving task was an investigation of the operation of a car jack—an investigation based on two isosceles triangles, where $AB = BC = BP$ (see fig. 6.4a). Anna and Kate conjectured that the car attachment point moved in a path perpendicular to the ground, but because of their experience with Tchebycheff's linkage, they were unsure whether the angle was exactly 90°. Tracing the path of C in the interactive geometry model (fig. 6.4b) and measuring $\angle CAP$ (fig. 6.4c) offered support for their conjecture and gave them the confidence to seek a geometric proof. Although this was the students' first attempt at constructing a proof (see fig. 6.5), their diagram and written argument display an understanding of the geometry of the linkage and of why $\angle CAP$ is a right angle.

Pascal's Angle Trisector

Pascal's angle trisector (see fig. 6.6), introduced to the students as "Pascal's mathematical machine" to disguise its purpose, offered more scope for data gathering. The linkage is again based on two isosceles triangles, with $AB = BC = DC$. Hinges occur at stationary points A and B, and bars BC and CD are hinged at C. As the bar AY rotates about point A, point C slides along AX and point D slides along AY. After dragging the interactive geometry linkage, Anna and Kate decided fairly quickly that it was angles, not lengths, that were likely to be significant.

Empirical evidence, conjecturing, and deductive reasoning were closely associated. Anna and Kate noted that triangles ABC and BCD were isosceles,

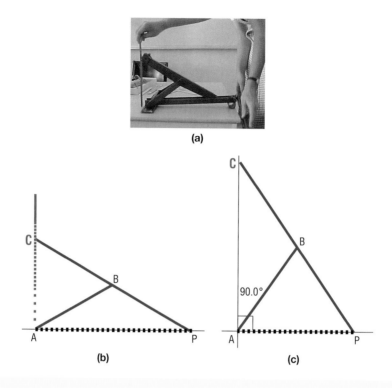

Fig. 6.4. Isosceles triangle car jack

and they recognized that the exterior angle, ∠*DBC*, was twice the size of ∠*BAC*. It was, however, the measurement and tabulation of angle sizes in the interactive geometry model (see fig. 6.7) that focused their attention on △*ADC* and enabled them to conjecture that there was a relationship among ∠*BAC*, ∠*ADC*, and ∠*DCX*.

> *Kate:* This angle, 83.7, equals 55.8 plus 27.9.
> *Teacher:* So what is your conjecture?
> *Kate:* So this angle [∠*DCX*] is equal to that [∠*BAC* = *a*] plus that [∠*ADC* = *b*], *a* plus *b*.
> *Anna:* So it equals 3*a*.

During their argumentation and conjecturing, Anna and Kate were encouraged by the teacher's intervention to justify verbally each statement they made so that when they came to their written proof constructions, justifications became a natural part of the process (see fig. 6.8).

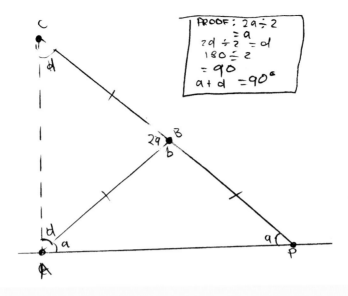

Fig. 6.5. Anna and Kate's "proof" for the car jack linkage

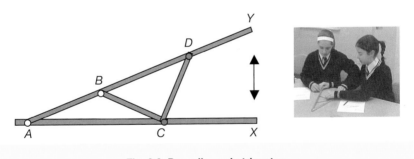

Fig. 6.6. Pascal's angle trisector

Sylvester's Pantograph

The students' use of the software was by no means restricted to dragging, and in fact it was their competence in the use of the software tools that allowed them to exploit its capabilities. In Sylvester's pantograph (see fig. 6.9), for example, the purpose of the device as a drawing instrument led naturally to the use of the software's Trace facility. Kate was quick to recognize the significance of the convergence of the traces of points P and P' (see fig. 6.9a) and excitedly exclaimed: "There! We've got them to meet!" followed by Anna's response: "Oh, we've got the angle there...." Since the software's

	∠YAX = ∠BCA	∠ABC = ∠CDY	∠DBC = ∠BDC	∠BCD	∠DCX
1	20.9	138.1	41.9	96.3	62.8
2	18.2	143.6	36.4	187.2	54.6
3	27.9	124.2	55.8	68.4	83.7
4					
5					

Note: The angle names in the column headings have been added for clarity and were not present in the students' screen tabulation.

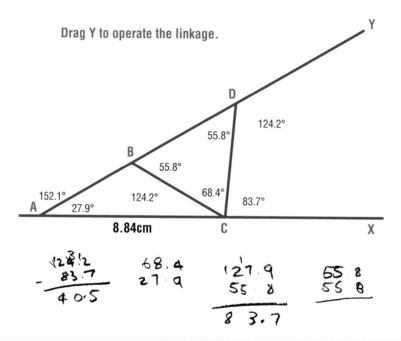

Fig. 6.7. Looking for angle relationships

Trace feature is transient, Anna and Kate drew segments over the convergent paths of *P* and *P'*, then they measured the angle between the two segments (see fig. 6.9b). They identified this angle as the angle of rotation of the image. The measurement of the fixed angles of the pantograph, ∠BAP ≅ ∠BCP', then supported their conjecture that the image was rotated through this angle (see fig. 6.9c). When prompted to observe the point about which the pantograph was rotating, Kate noted that *OP* and *OP'* were equal. Measuring ∠POP' led the students to the conjecture that ∠POP' represented the angle of rotation and that ∠POP' ≅ ∠BAP ≅ ∠BCP'.

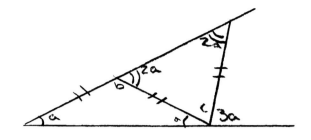

Given: $AB = BC = CD$

Prove: $\angle DCX = \angle BAC \times 3$

Proof: $a = \angle BAC$ and $\angle BCA$

$b = \angle ABC$ ∴ $\angle CBD = 2a$ (exterior angle)

∴ $\angle ADC = 2a$ (isoceles triangle)

∴ $\angle DCX = 3a$ (exterior angle)

Fig. 6.8. Anna and Kate's diagram and proof for Pascal's angle trisector

Using their knowledge of rhombus properties, Anna and Kate were able to deduce the angle relationships in the pantograph (see fig. 6.10) and to use these relationships in constructing their proof (see fig. 6.11).

Another pair of students—Lucy and Rose—commenced their investigation of the interactive geometry pantograph by drawing a triangle and dragging P around it so that the image was formed by the trace of P'. They constructed a triangle over the image trace (fig. 6.12) and carefully dragged the original triangle, placing it over the image triangle in order to measure the angle of rotation between the two triangles. As with Anna and Kate, seeing the linkage as an accurate geometric diagram encouraged Lucy and Rose to add construction lines and to notice congruent angles. Measuring $\angle PAB$ and $\angle P'CB$ then led Lucy and Rose to the same conjecture—that the angle of rotation was equal to the fixed angle of the pantograph.

As a result of the strong empirical support from the software and the argumentation that accompanied the production of each conjecture, Lucy and Rose already had a strong sense of logical order for the steps of deductive reasoning when they came to construct their written proof, as illustrated by Rose's suggestion: "Let's do the sides first. OA equals AP equals OC equals CP' ... then angle OCP' equals ... OAP because they both have 30 degrees ... they share 30 degrees ... we shouldn't do that yet ... angle OA ... angle OAB equals ..."

Rose's subsequent comment—"Once we've proved that angle, then the whole thing's easy 'cause side angle side ... see, if you have two sides and

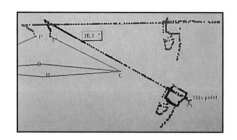

(a)

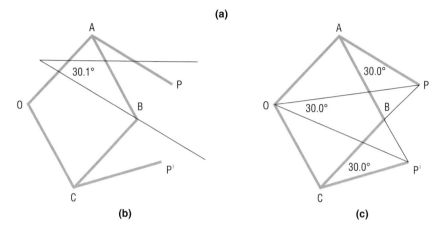

(b)   **(c)**

Fig. 6.9. Exploring the interactive geometry model of Sylvester's pantograph

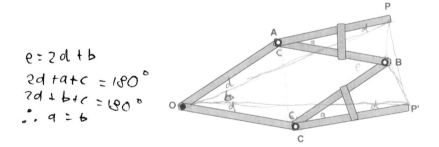

$$e = 2d + b$$
$$2d + a + c = 180°$$
$$2d + b + c = 180°$$
$$\therefore a = b$$

Fig. 6.10. Identifying angle relationships in Sylvester's pantograph

Prove: $\angle POP' = \angle BCP' = \angle PAB$
Proof: $\angle COP' = \angle CP'O$ (isoseles triangle)
$= \angle AOP = \angle OPA$ (congruent triangles) $= d$
let angle $OCB = c$
$d + d + c + a = 180°$ (triangle)
$\angle AOC + \angle OCB = 180°$ (rhombus), let angle $POP' = b$
∴ $d + d + c + b = 180°$
∴ $b = a$
∴ $\angle POP' = \angle BCP' = \angle PAB$

Fig. 6.11. Anna's proof for the angle of rotation

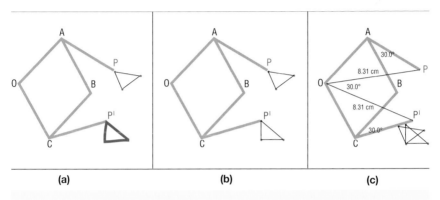

(a) **(b)** **(c)**

Fig. 6.12. Lucy and Rose: Determining the angle of rotation

how big it's going to be in between … when you join them up, the triangles will be the same"—suggests that she was able to engage in the sort of reasoning referred to by Simon (1996) as transformational reasoning. It would appear that Rose's reasoning may have been influenced by the dynamic visualization in the computer environment, lending support to the assertion by Scher (1999, p. 24) that interactive geometry software can influence the style of experimentation and reasoning so that "the boundary between deductive reasoning and dynamic geometry becomes blurred: the software finds its way into the proof process."

Consul, the Educated Monkey

The most complex of the linkages investigated by the students was "Consul, the Educated Monkey" (see fig. 6.13a; Kolpas and Massion 2000)— a 1916 American mathematical toy designed to teach multiplication tables. Figure 6.13 shows (a) a geometric diagram of the mechanical linkage superimposed on Consul, (b) geostrip models of the Consul linkage, and (c) an

(a) Consul with super-imposed linkage

(b) Plastic geostrip model

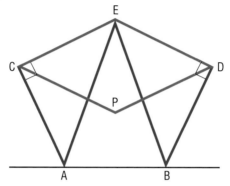

(c) Interactive geometry model of the linkage

Fig. 6.13. Consul, the educated monkey

interactive geometry simulation of the linkage. Each upper arm and leg is constructed from a single piece of tin plate and pivots about a point *(E)* beneath the bow tie, with the slotted tail ensuring that E and P move vertically. Segments *CE, DE, CP,* and *DP* (as well as the distances *AC* and *BD*) are all equal, so that *CEDP* is a rhombus that is pivoted at points *C* and *D* (the monkey's elbows) and points *E* and *P.* Angles *ACE* and *BDE* are right angles, so that angles *CEA* and *DEB* remain constant and equal to $45°$. The feet, at *A* and *B*, can move only in a straight line through *A* and *B*, and when positioned on any pair of factors, *P* points to their product.

Anna and Kate moved backward and forward among the toy calculator, the geostrip model, and the computer model in their attempt to understand the geometry of the linkage. Tracing the path of *P* in the interactive geometry model (see fig. 6.14a) led them to conjecture that the operation of the toy calculator depended on ∠*APB* remaining a right angle.

Kate: See, it [*P*] goes along there … [*points to PA*].

Anna: And then if we do it this way … if we move *A*, it'll go down the other way.

Further angle measurements (see fig. 6.14b) enabled Anna and Kate to conjecture that $\triangle AEB$ and $\triangle ACP$ were similar.

Laborde (1998) asserts that dynamic drawings offer stronger visual evidence than a single static drawing: "A spatial property may emerge as an invariant in the movement whereas this might not be noticeable in one static drawing" (p. 117). She notes that when students are engaged in problem-solving tasks in interactive geometry computer environments, "a critical point of the solving process is the visual recognition of a geometrical invariant by the students, which allows them to move to geometry" (p. 120).

Anna and Kate's sustained reasoning during their proof construction and the ease with which they were able to order their statements logically provide strong support for the notion of cognitive unity proposed by Boero et al. (1996). Boero and his colleagues believe that the conjectures and associated justifications put forward during the argumentation play a principal role in the logical ordering of statements and are essential to students' successful construction of proofs.

Fig. 6.15 shows Anna and Kate's written proof for the Consul linkage. Although initially it was not an intention of the current research to assess the students' argumentations quantitatively, the model developed by Galindo (1998) to assess students' reasoning in an interactive geometry environment was found to be useful in comparing the extent of interaction between the students' empirical and deductive justification. Galindo asserts that there should be an empirical component and a deductive reasoning component to students' explorations, with interaction between the two components. His assessment model is therefore based on three levels of justification: *intuitive justification, deductive justification,* and *interplay between intuitive and deductive,* with three scores within each category: 0 (no justification or no evidence of interplay), 1 (partial justification or some evidence of interplay), or 2 (reasonable justification or mutually reinforcing justification). Galindo notes that the desired score for a particular task would depend on the focus of the task, but "when students are expected to make explicit the connections they are making between the empirical and deductive bases of their reasoning, the goal should be to obtain 2-2-2 scores" (p. 81). The argumentations of the van Hiele Level 2–3 students, such as Anna and Kate, and Lucy and Rose, satisfied the criteria for 2-2-2 scores. The Level 1–2 students, although often relying substantially on the teacher's intervention during their argu-

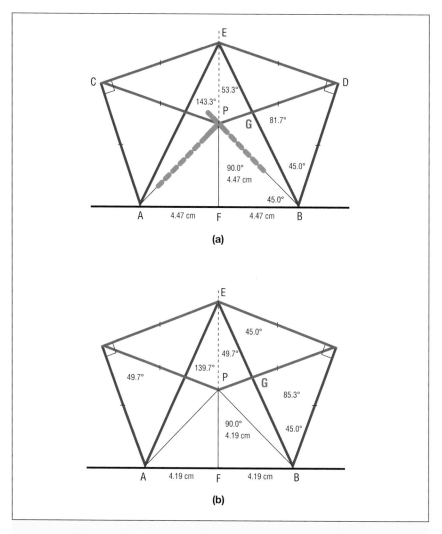

Fig. 6.14. Interactive geometry feedback supports conjecturing

mentations, were nevertheless able to engage in software-supported deductive reasoning. However, the argumentations of these students typically would be assessed as 2-1-1 or 2-1-2.

Discussion

The motivating context and the dynamic visualization associated with the linkages fostered conjecturing and intense argumentation. There seemed little

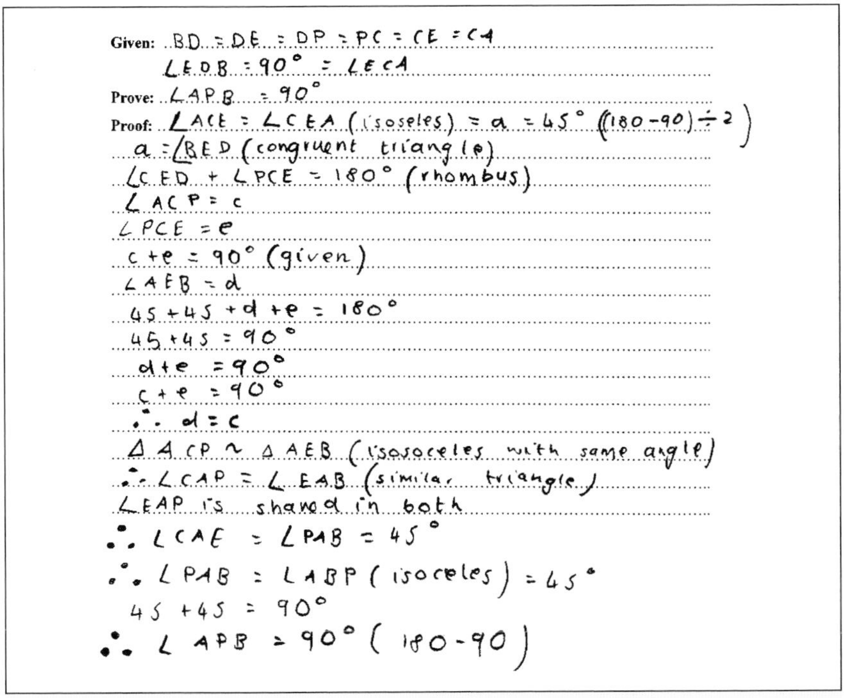

Given: $BD = DE = DP = PC = CE = CA$
$\angle EDB = 90° = \angle ECA$

Prove: $\angle APB = 90°$

Proof: $\angle ACE = \angle CEA$ (isoseles) $= a = 45°$ $\left((180-90) \div 2\right)$
$a = \angle BED$ (congruent triangle)
$\angle CED + \angle PCE = 180°$ (rhombus)
$\angle ACP = c$
$\angle PCE = e$
$c + e = 90°$ (given)
$\angle AEB = d$
$45 + 45 + d + e = 180°$
$45 + 45 = 90°$
$d + e = 90°$
$c + e = 90°$
$\therefore d = c$
$\triangle ACP \sim \triangle AEB$ (isosoceles with same angle)
$\therefore \angle CAP = \angle EAB$ (similar triangle)
$\angle EAP$ is shared in both
$\therefore \angle CAE = \angle PAB = 45°$
$\therefore \angle PAB = \angle ABP$ (isoceles) $= 45°$
$45 + 45 = 90°$
$\therefore \angle APB = 90°$ $(180 - 90)$

Fig. 6.15. Anna and Kate's proof for the Consul linkage

doubt that the majority of students enjoyed the tactile experience of operating the linkages, and their curiosity was aroused to find out why each linkage worked in the observed way. Indicators of motivation (Helme and Clarke 2001)—enjoyment, excitement, satisfaction, and perseverance—and of cognitive engagement—reflective statements, statements that involved shared meaning, or statements that drew on earlier interactions—were apparent throughout the argumentations of the students.

The underlying geometric shapes on which the linkages were based were sufficiently familiar for all students to be able to engage in productive argumentation. The teacher's intervention was an important feature of the students' argumentations—prompting the students to furnish justifications for their statements and checking the validity of their justifications. In a relatively short period of time, the van Hiele Level 2–3 students, as exemplified by Anna and Kate, had made remarkable progress in their ability to reason deductively and to order their statements logically in written proofs. Even those students who commenced at Levels 1–2, although more reliant on the teacher's support, developed an understanding of the requirements of deductive reasoning and made considerable progress toward Level 2 and 3 understanding.

Although the tactile experience and satisfaction of working with real (mechanical) and geostrip linkages represented a significant motivational aspect, the students recognized that the computer models offered them more useful empirical feedback. Their trust in the interactive geometry data strengthened their confidence in their conjectures and encouraged them to seek geometric explanations. The acceptance by these Year 8 students of the need for proof and their sustained engagement in the process of proof construction, despite the strength of conviction based on empirical interactive geometry data, substantiates the claim by Hölzl that it is the context, not the computer itself, that determines whether or not a need for proof arises. Hölzl asserts that it is the quest for explanation that drives the reasoning process, which indeed seemed to be so for these Year 8 students, as illustrated by Anna and Kate's investigation of the Consul linkage:

> *Kate:* I just want to draw this triangle.
>
> *Anna:* You're right. It goes down the triangle … see, this point P goes down … it goes down … like along the triangle …

Bartolini Bussi and Pergola (2000, p. 61), referring to an investigation of Sylvester's pantograph by Year 11 students, note that "the linkage itself created the need to be understood."

In the context of these tasks, proof assumed the multiple roles of a *verification* of the truth of conjectures, an *understanding* of geometric relationships, and an *explanation*, that is, giving insight into *why* a particular linkage works in the observed way. Once their curiosity had been aroused through the linkage tasks, the students approached pencil-and-paper proofs and open-ended interactive geometry investigations with the same enthusiasm, accepting the challenge to construct valid geometric arguments for their conjectures.

REFERENCES

Australian Education Council. *A National Statement on Mathematics for Australian Schools.* Carlton, Victoria, Australia: Curriculum Corp., 1991.

Bartolini Bussi, Maria G., and Marcello Pergola. "History in the Mathematics Classroom: Linkages and Kinematic Geometry." In *History of Mathematics and Education: Ideas and Experiences,* edited by H. Niels Jahnke, Norbert Knoche, and Michael Otte, pp. 39–67. Göttingen, Germany: Vandenhoeck & Ruprecht, 2000.

Boero, Paolo, Rossella Garuti, Enrica Lemut, and M. Alessandra Mariotti. "Challenging the Traditional School Approach to Theorems: A Hypothesis about the Cognitive Unity of Theorems." In *Proceedings of the 20th Conference of the International Group for the Psychology of Mathematics Education,* edited by Luis Puig and Angel Gutiérrez, vol. 2, pp. 113–20. Valencia, Spain: PME, 1996.

Bolt, Brian. *Mathematics Meets Technology.* Cambridge: Cambridge University Press, 1991.

Galindo, Enrique. "Assessing Justification and Proof in Geometry Classes Taught Using Dynamic Software." *Mathematics Teacher* 91 (January 1998): 76–82.

Goldenberg, E. Paul, Al Cuoco, and June Mark. "A Role for Geometry in General Education." In *Designing Learning Environments for Developing Understanding of Geometry and Space,* edited by Richard Lehrer and Daniel Chazan, pp. 3–44. Mahwah, N.J.: Lawrence Erlbaum, 1998.

Hanna, Gila, and H. Niels Jahnke. "Proof and Application." *Educational Studies in Mathematics* 24 (1993): 421–38.

Helme, Sue, and David Clarke. "Cognitive Engagement in the Mathematics Classroom." In *Perspectives on Practice and Meaning in Mathematics and Science Classrooms,* edited by David Clarke, pp. 131–53. Dordrecht, Netherlands: Kluwer Academic Publishers, 2001.

Hölzl, Reinhard. "Using Dynamic Geometry Software to Add Contrast to Geometric Situations—a Case Study." *International Journal of Computers for Mathematical Learning* 6 (2001): 63–86.

Kolpas, Sidney J., and Gary R. Massion. "Consul, the Educated Monkey." *Mathematics Teacher* 93 (April 2000): 276–79.

Laborde, Colette. "Visual Phenomena in the Teaching/Learning of Geometry in a Computer-Based Environment." In *Perspectives on the Teaching of Geometry for the 21st Century,* edited by Carmelo Mammana and Vinicio Villani, pp. 113–21. Dordrecht, Netherlands: Kluwer Academic Publishers, 1998.

Lawrie, Christine J. "An Investigation into the Assessment of a Student's Van Hiele Levels of Understanding in Geometry." Ph.D. diss., University of New England, Armidale, New South Wales, Australia, 1997.

National Council of Teachers of Mathematics (NCTM). *Principles and Standards for School Mathematics.* Reston, Va.: NCTM, 2000.

Noss, Richard, and Celia Hoyles. *Windows on Mathematical Meanings: Learning Cultures and Computers.* Dordrecht, Netherlands: Kluwer Academic Publishers, 1996.

Scher, Daniel. "Problem Solving and Proof in the Age of Dynamic Geometry." *Micromath* 15 (spring 1999): 24–30.

Simon, Martin. "Beyond Inductive and Deductive Reasoning: The Search for a Sense of Knowing." *Educational Studies in Mathematics* 30 (1996): 197–210.

Vincent, Jill, Helen Chick, and Barry McCrae. "Mechanical Linkages as Bridges to Deductive Reasoning: A Comparison of Two Environments." In *Proceedings of the 26th Conference of the International Group for the Psychology of Mathematics Education,* vol. 4, edited by Anne D. Cockburn and Elena Nardi, pp. 313–20. Norwich, U.K.: PME, 2002.

Vincent, Jill, and Barry McCrae. "Mechanical Linkages and the Need for Proof in Secondary School Geometry." In *Proceedings of the 25th Conference of the International Group for the Psychology of Mathematics Education,* vol. 4, edited by Marja van den Heuvel-Panhuizen, pp. 367–74. Utrecht, Netherlands: PME, 2001.

7

Square or Not? Assessing Constructions in an Interactive Geometry Software Environment

Daniel Scher

C LASSROOM geometry has gone "dynamic." A square that can be resized with a click and drag of a computer mouse holds enormous appeal for a generation accustomed to the static, hands-off nature of textbook illustrations. Activities incorporating interactive geometry software such as The Geometer's Sketchpad (Jackiw 2001) and Cabri Geometry (Laborde and Bellemain 1994) appear on a nearly monthly basis in the *Mathematics Teacher,* one of the classroom journals of the National Council of Teachers of Mathematics (NCTM). In 1997, the Mathematical Association of America published *Geometry Turned On* (King and Schattschneider 1997), an entire volume devoted to applications of interactive geometry software. Several secondary school geometry curricula include computer explorations in their texts (Gay 1998; Serra 2003), and *Principles and Standards for School Mathematics* (NCTM 2000) recommends the use of the software to promote mathematical investigations.

This interest in motion geometry is not new. Syer, writing in 1945, describes the ability of film to create "continuous" geometric images. His advocacy of the moving picture reads much as a modern-day justification for interactive geometry software (Syer 1945, p. 344):

> In addition to true-to-life demonstrations of solid geometry, it would be interesting to make greater use of the peculiar advantages of moving pictures over ordinary models. In plane geometry films we used figures which changed shape, position, and color without distracting pauses or outside aid. This continuous and swift succession of illustrations is fast enough to keep

This study was supported by the National Science Foundation (NSF) grant RED-9453864 as part of the Epistemology of Dynamic Geometry Project at Education Development Center, Inc., in Newton, Massachusetts. The opinions expressed in this article are not necessarily those of the NSF. Thanks to E. Paul Goldenberg, Kenneth Goldberg, Fran Curcio, and Joe McDonald for their input.

Editor's note: The CD accompanying this yearbook contains one activity sheet and three Geometer's Sketchpad files that are relevant to this article. For more information on The Geometer's Sketchpad, contact Key Curriculum Press at www.keypress.com/.

up with a spoken description, or even as fast as the thought processes that are developing the idea. Thus no time is lost erasing pictures from the blackboard, changing lantern slides, or holding up illustrations, because the illustrations and thought move simultaneously.

Today's software incorporates the motion available in film but goes one step further, providing users the tools to design their own animations. Exercising such control over a moving image places new demands on students, since they must learn how to build constructions that respond appropriately when dragged with the computer mouse. A well-behaved square, for example, will change its size and orientation when dragged, but not its shape.

How students negotiate the tools of interactive geometry software to accomplish their construction goals remains an open question in mathematics education literature (Goldenberg 1998). Are there instances where a particular software tool or technique impedes a student from completing a construction task (Laborde 1993)? Or do the software's tools promote novel construction techniques that would not surface in a straightedge-and-compass environment?

This article describes a student-interview study conducted with E. Paul Goldenberg of Education Development Center that focused on the learning of geometry in an interactive geometry software environment. Middle school students received prebuilt "mystery" constructions on The Geometer's Sketchpad that included such common geometric objects as squares, rectangles, isosceles triangles, and perpendicular bisector lines. Interviewees explored these objects by dragging each of their parts with the computer mouse (for a collection of similar activities, see Battista 2003).

As they experimented, students described what they observed on the screen and explained how they thought each object was built. Beginning then with a fresh blank screen, students attempted to reconstruct the identical objects from scratch. Throughout the interview, videotape recorded the mouse movements and menu selections of the students, as well as their accompanying commentary. The detailed nature of the tapes makes them an ideal source for analyzing the geometric conceptions of students in an interactive geometry setting.

The Interview Setting and Procedures

Our interview study targeted middle school students with no prior interactive geometry experience. In total, we interviewed three sixth graders and five seventh graders from two middle schools (six boys and two girls). Each student participated in two interview sessions held on separate days, with individual sessions running approximately two hours. As the students explored and performed constructions with the software, one camera recorded the computer screen while another camera videotaped the interviewee. These separate images were transferred to a mixing board sitting outside the interview room, where a technician combined the shots to produce a split-screen composite tape.

As the students progressed through the interview tasks, the interviewer sat by their side and functioned in two roles:

1. When a student was uncertain whether the software contained a particular feature or forgot where it was located, the interviewer offered assistance. Throughout the session, the interviewer reminded students that the interview was not a test of how well they had memorized the software commands; rather, it was intended to uncover how they thought about the objects on screen.

2. From time to time, the interviewer would ask questions like, "What are you trying to do? Describe to me what you're seeing. Can you explain why that line behaves the way it does? How might you test your theory?" The interviewer would also restate or rephrase some of the students' observations to spotlight ideas that would benefit from further attention.

Constructing a Square

Compared to other geometric shapes, a square is a relatively simple object to construct with interactive geometry software, built with a circle tool and repeated applications of a perpendicular line command. The interview excerpts that follow, however, offer students construction techniques that depart from standard methods. With each technique comes the same question: Have the interviewees built a square? Although this may sound like a simple matter to answer, it is, in fact, a thornier issue than one might suppose in the world of interactive geometry.

Given a quadrilateral drawn on a piece of paper, it is easy enough to check if it fits the definition of a square—measure its sides and angles. If the sides are equal and the angles measure 90 degrees, the quadrilateral is a square. By contrast, an interactive geometry quadrilateral might pass these measurement tests and still not qualify as a square. To clarify the situation, Finzer and Bennett (1995) established four categories for describing or judging the merits of a figure built with interactive geometry software. A square, or other object, is either a Drawing, Underconstrained, Overconstrained, or Appropriately Constrained. Figure 7.1 shows a square that reveals itself to be a *drawing* when any vertex is dragged (see the sketch called "Draw vs. Construct.gsp" on the CD accompanying this yearbook). In this instance, the supposed square was created by eyeballing and then measuring the lengths and angles formed by four segments so they would appear equal in length and posi-

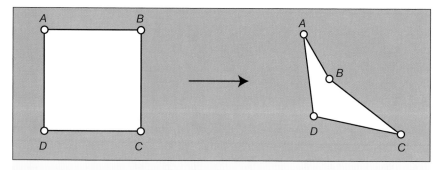

Fig. 7.1. *ABCD* looks like a square, but it isn't.

tioned at right angles. Without any geometric constraints built into the picture, any perturbation of the object deforms it into an arbitrary quadrilateral.

Figure 7.2 shows an *underconstrained* square. It has four built-in right angles but no constraints to keep its lengths equal. Dragging vertex *B* deforms the square into a rectangle.

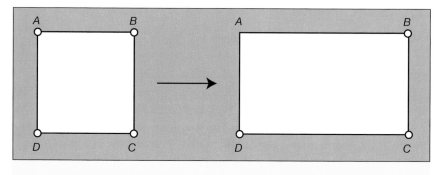

Fig. 7.2. *ABCD* looks like a square, but it is actually a rectangle.

In contrast to an underconstrained square, an *overconstrained* square does retain four equal sides when dragged. These sides, however, remain steadfastly fixed in length and thus depict only a single square.

Finally, figure 7.3 shows a common method for building a bona fide square with *appropriate constraints*. The construction begins by drawing a circle with center *D* and a point *A* on its circumference. Point *A*, known as a "control point," changes the size of the circle when dragged, as does point *D*. By first drawing segment *DA* and then using a perpendicular command, the user can construct a segment *DC* of equal length. In a similar manner, segments

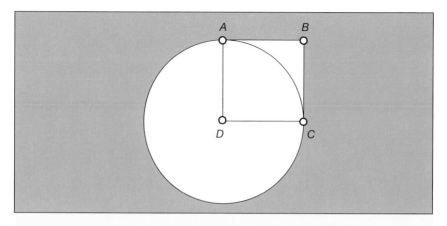

Fig. 7.3. Constructing an appropriately constrained square

AB and *BC* are constructed perpendicular to segments *DA* and *DC* respectively, guaranteeing that all four of the quadrilateral's angles measure 90 degrees. With these construction features in place, *ABCD* may grow, shrink, or rotate when tugged but will always remain square.

The remainder of this article highlights several interviewees' square-building efforts. Their methods expand on the construction distinctions above and raise new questions concerning the definition of an interactive geometry square.

Norman's Square

When Norman begins construction on his square, the interviewer tells him that he'll first need to "learn some things" about the software's menu items (such as the Perpendicular Line command). Undaunted, Norman assures him, "No, you can do it in a much easier way."

Norman draws a square *ABCD* by freehand, carefully estimating the positions and lengths of the four on-screen segments. He then selects the software's circle tool and draws a circle that originates roughly in the center of *ABCD* and extends out to its four vertices (see fig. 7.4 and the sketch called "Norman.gsp" on the CD accompanying this yearbook). Norman draws the circle so that point *A* serves as its control point.

Norman drags point *A* and admits that his square has some problems. Since its lengths and angles were estimated (or drawn) rather than constructed, *ABCD* deforms into a random quadrilateral. Judged by Finzer and Bennett's classification, Norman has himself a drawing.

Yet beneath this seeming failure lies the germ of a good idea. When a circle expands, it retains its shape. If *ABCD* can be linked to the circle, perhaps it, too,

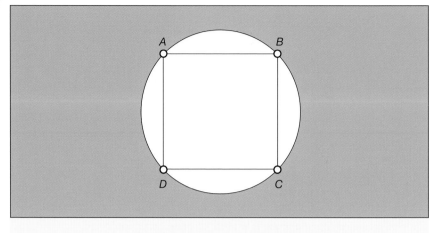

Fig. 7.4. Norman's initial square-building method

will retain its shape. Because Norman drew his circle *after* forming *ABCD*, none of the points *B*, *C*, and *D* remains attached to its circumference. Norman wants all four points of *ABCD* to move in unison with the circle, so he reverses course.

Norman begins afresh by drawing another circle. Only then does he draw *ABCD*, placing each of its four vertices along the circle by visual inspection. Again, point *A* serves as the circle's control point. When Norman drags point *A* of his new square randomly across the screen, the behavior of *ABCD* exhibits more regularity than his previous attempt. Now, all four points of the quadrilateral remain attached to the circle when it moves. Aside from segments *AB* and *AD*, the entire figure grows and shrinks proportionately. In other words, $BC = CD$ and $\angle BCD = 90$ degrees regardless of the quadrilateral's size (see fig. 7.5). Some careful dragging of point *A* yields more regularity still: if point *A* is dragged in a northwest direction (keeping the measure of $\angle DAB$ equal to 90 degrees), *ABCD* remains a square while expanding.

Commentary

Norman's square-building technique stays clear of the perpendicular and parallel line construction tools. Without the use of these menu items, it is hard to imagine how anything he builds could rise above the category of drawing. Yet through the clever use of a circle, he manages to build a quadrilateral *ABCD* that grows and shrinks, all the while approximating a square, provided point *A* moves in a controlled path.

Norman's efforts fall short of an *appropriately constrained* object, but there is still much to admire in his work:

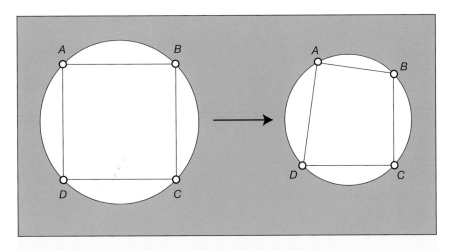

Fig. 7.5. The result of dragging point A

1. *Economy and speed.* Building figure 7.3's square demands multiple trips to the software's Perpendicular or Parallel Line commands. It also requires the user to hide construction lines and replace them by segments. By contrast, Norman's method involves no hidden lines and no menu selections. It can be completed in under a minute, qualifying it as a quick-and-simple means of illustrating a multitude of squares.

2. *Mathematical integrity.* None of the construction elements built into figure 7.1's drawing of a square help to keep its four sides equal in length. By contrast, Norman's method uses the symmetry of a circle to achieve four (sometimes) equal lengths.

Because Norman's construction exhibits "square" behavior, one might classify it an *underconstrained* square. This assessment, however, leaves room for disagreement. Norman drew his square by eyeballing the correct locations for its vertices. As such, the angle measures of *ABCD* were not precisely 90 degrees, and its lengths were not perfectly equal. These inaccuracies can be corrected by measuring and adjusting the particulars of *ABCD*, but is measuring a legitimate part of the construction process? Traditionalists would likely say no.

Regardless of which side of this debate one chooses to accept, the message here is more general: deciding whether an object built with interactive geometry software is a square, contains some degree of "squareness," or is not a square can be surprisingly difficult and open to multiple perspectives.

David and Ben's Square

David and Ben begin their square construction by drawing a circle and placing points B and D on its circumference (see the sketch called David and Ben.gsp on the CD accompanying this yearbook). They add radii $\overline{CB}$ and $\overline{CD}$ to the picture and construct perpendicular lines j and k through points B and D, respectively (see fig. 7.6a).

David and Ben's construction is similar to the appropriately constrained square in figure 7.3, but with one difference: whereas $\angle BAD$ in figure 7.3 always measures 90 degrees, $\angle BAD$ in figure 7.6 can assume a variety of measurements, since points B and D move independently. Only when $\overline{CB}$ is perpendicular to $\overline{CD}$ is $ABCD$ square (fig. 7.6b).

The interviewer questions David and Ben about $ABCD$ but soon discovers that neither interviewee intends this quadrilateral to serve as his final square:

> *Int.:* You're trying to build a square. What are the features you're trying to build into it?
>
> *Ben:* Equal sides and ninety-degree angles, all four.
>
> *Int.:* Check this thing out by moving it in various ways. And find out whether any of the properties that you're looking for are

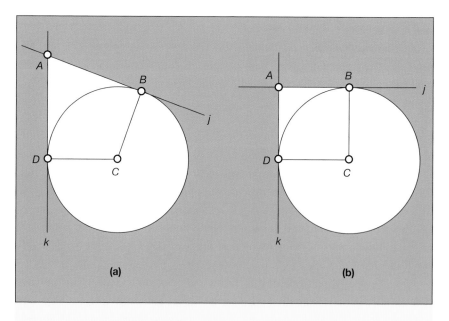

Fig. 7.6. David and Ben's construction method

there. For example, you do have some ninety-degree angles that stay all the time.

Ben: Well, we don't have a full square yet so we can't really tell.

David: Yeah we do. That's a full square [*points to ABCD*].

Ben: Oh yeah. I wasn't looking at that part.

David: Actually, I wasn't really thinking of that either. I was thinking we'd go all the way around it [the circle].

Ben: Yeah, I was too.

David and Ben plan to build a square that circumscribes (goes "all the way around") their circle. They resume their construction by placing points E and F on the circle's circumference, drawing $\overline{CE}$ and $\overline{CF}$, and constructing lines perpendicular to the two segments (see fig. 7.7a).

By dragging points E and F to their proper locations, David and Ben are able to make their quadrilateral look like a square (see fig. 7.7b). Of course, any subsequent dragging of $D, B, E,$ or F will deform the square back into an arbitrary quadrilateral.

At this point, the interviewer intervenes, explaining that since segments BF and DE were *drawn* (as opposed to *constructed*) perpendicular to each other, the illustration will never be an appropriately constrained square. Not to be deterred by this observation, David suggests a simple fix:

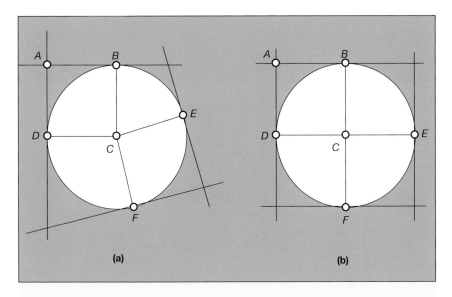

(a)   (b)

Fig. 7.7. David and Ben continue their square construction.

David: Wait, wait, wait. I have a question. If you hide points *D, A, B, E,* and *F,* and you could only move point *C,* would that work?

Sure enough, when David hides all points except the circle's center (fig. 7.8a) and then drags point *C,* their quadrilateral grows and shrinks, always remaining a square regardless of point *C*'s location (fig. 7.8b). David and Ben give each other a congratulatory high-five and move on to the next challenge.

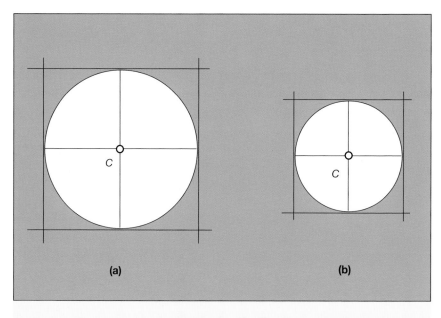

(a)   (b)

Fig. 7.8. Hiding every point except *C* yields a resizable square.

Commentary

As soon as David and Ben placed independent points *B* and *D* on their circle (fig. 7.6), I was tempted to curtail their work. My experience with appropriately constrained squares told me there was too much variability in their construction. David and Ben intended angle *BCD* to measure 90 degrees, but I knew a simple tug of either point *B* or *D* would alter the angle's measure.

David and Ben's solution to this problem—hiding nearly every point in their sketch (fig. 7.8)—struck me at first as cheating. Yes, dragging point *C* did enlarge and shrink their square perfectly well, but the hidden points *B, D, E,* and *F* could "mess up" the shape. Was it fair to remove these points from view, thus making them inaccessible to dragging by the mouse? Strictly speaking, yes. Hiding extraneous construction points is perfectly legal under the laws of interactive geometry constructions. Yet if David and Ben had *constructed* (as opposed to *drawn*) the segments perpendicular to each other, then there would be no need to hide points. Any point, visible or otherwise, would maintain the quadrilateral's square shape when dragged.

Conclusion

In the language of interactive geometry, a square is created as either a drawing or a construction. To be more specific still, a square can be categorized as a drawing, underconstrained, overconstrained, or appropriately constrained. Theoretically, the differences among these four Finzer and Bennett (1995) categories create an unambiguous scheme for classifying students' construction attempts. In practice, these distinctions proved decidedly murky. Students' efforts at building geometric shapes from scratch sometimes defied categorization as "right" or "wrong." The prevailing classification structure, although helpful, did not account for creative construction techniques.

Norman, David, and Ben built squares with minimal, if any, use of common interactive geometry software tools like "perpendicular line." Through clever application of a circle's symmetry, these students were able to build squares that, to varying degrees, maintained their "squareness" when dragged. Such nontraditional approaches serve as a challenge to teachers who assess the construction efforts of their students.

Students' work may not fit strict construction standards, yet they may still contain nuggets of mathematical insight. If a student builds a square that sometimes, but not always, maintains its squareness (and is thus "underconstrained"), should she be commended for her work? I would argue yes. Although the quadrilateral lacks "appropriate constraints," it is no small achievement to elevate a square above the category of "a drawing." Teachers must look freshly at each of their students' interactive geometry constructions and not adopt a lockstep appraisal method.

If several students in a class build squares using different techniques, a teacher can use this opportunity to ask them to compare and contrast the merits of each construction. What characteristics of each object qualify it as a square? Where does each square fall short? Can the class form a consensus on how an interactive geometry square should behave? Student-led critiques may be more effective than teacher-imposed definitions.

REFERENCES

Battista, Michael. *Shape Makers: Developing Geometric Reasoning with The Geometer's Sketchpad.* Emeryville, Calif.: Key Curriculum Press, 2003.

Finzer, William F., and Dan S. Bennett. "From Drawing to Construction with The Geometer's Sketchpad." *Mathematics Teacher* 88 (May 1995): 428–31.

Gay, David. *Geometry by Discovery.* New York: John Wiley & Sons, 1998.

Goldenberg, E. Paul. "What Is Dynamic Geometry?" In *Designing Learning Environments for Developing Understanding of Geometry and Space,* edited by Richard Lehrer and Dan Chazan, pp. 351–67. Mahwah, N.J.: Lawrence Erlbaum Associates, 1998.

Jackiw, Nicholas. The Geometer's Sketchpad. Software. Ver. 4.0. Emeryville, Calif.: KCP Technologies, 2001.

King, James, and Doris Schattschneider, eds. *Geometry Turned On.* Washington, D.C.: Mathematical Association of America, 1997.

Laborde, Colette. "The Computer as Part of the Learning Environment: The Case of Geometry." In *Learning from Computers: Mathematics Education and Technology,* edited by Christine Keitel and Kenneth Ruthven, pp. 48–67. Berlin and New York: Springer-Verlag, 1993.

Laborde, Jean-Marie, and Franck Bellemain. Cabri Geometry II. Software. Dallas, Tex.: Texas Instruments, 1994.

National Council of Teachers of Mathematics (NCTM). *Principles and Standards for School Mathematics.* Reston, Va.: NCTM, 2000.

Serra, Michael. *Discovering Geometry: An Investigative Approach.* Emeryville, Calif.: Key Curriculum Press, 2003.

Syer, Henry W. "Making and Using Motion Pictures for the Teaching of Mathematics." In *Multi-Sensory Aids in the Teaching of Mathematics,* Eighteenth Yearbook of the National Council of Teachers of Mathematics, edited by W. D. Reeve, pp. 325–45. New York: Bureau of Publications, Teachers College, Columbia University, 1945.

8

An Anthropological Account of the Emergence of Mathematical Proof and Related Processes in Technology-Based Environments

F. D. Rivera

Man's use of mind is dependent upon his ability to develop and use "tools" or "instruments" or "technologies" that make it possible to express and amplify his powers.... It was not a large-brained hominid that developed the technical-social life of the human; rather it was the tool-using, cooperative pattern that gradually changed man's morphology.... What evolved as a human nervous system was something, then, that required outside devices for expressing its potential.

—Jerome S. Bruner, *Toward a Theory of Instruction*

CURRENT research on mathematical thinking and understanding has produced important information about how learners construct knowledge. However, the psychological orientation of many of these studies somehow downplays the instrumental role that tools play in cognitive processing despite the anthropological fact that all human experience is mediated by and structured through material, semiotic, and technological systems. In this paper, *mathematical processes in technology-based environments* refer to context-specific operations and actions that learners exhibit as they develop facility with different functions of technological tools that have been purposefully constructed to extend and amplify human capabilities. Thus, both epistemological and pragmatic concerns about the role of technological tools in the development of mathematical processes address the relationship between and among learner, tool, and task (see fig. 8.1), including the individual and social transformations that emerge from the interaction.

Editor's note: The CD accompanying this yearbook contains six Cabri activities that are relevant to this article. For more information on Cabri, contact Cabri at www-Cabri.imag.fr/Cabri/index-e.html or at education.ti.com/us/product/software/cabri/features/features.html.

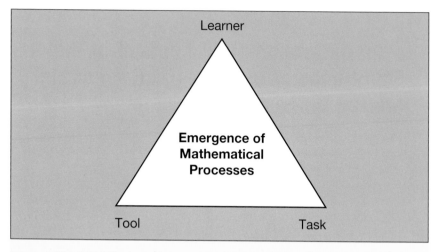

Fig. 8.1. Anthropological perspective on the emergence of mathematical processes

A number of important anthropological research studies have explored the problem concerning how technological tools aid in the emergence of proof and related mathematical processes such as dynamic visualization. At the outset, these studies posit the view that growth in mathematical knowledge is mediated by tools that determine individual (mental) and social (collaborative) action. The anthropological standpoint breaks away from psychological views that often frame learners as developing attributes of, or relationships within, mathematical objects independent of the tools that aid their thinking process. Instead, it focuses on the microgenetic interaction—by way of *instrumental genesis*—that takes place between learners and the tools they use that help them acquire knowledge about some task being explored. For instance, a typical constructivist-based research study investigates ways in which middle school students establish the property that the sum of the interior angles in any triangle is 180° by some process (i.e., internally (radical) or through collective argumentation (social)) within a learning activity. The anthropological route, however, investigates ways in which learners manipulate tools that they use to perform the activity, and it then leads to the identification of tool-induced processes that help them acquire the angle-sum property.

Thus, from an anthropological perspective, the didactic function of tools in a mathematical activity is seen as moving across a mere artifactual view of tools—that is to say, in the literal sense as external objects that could do routine, mechanized computations—to the important mediating role they play in the development of (stable) mathematical processes. Recent anthropological studies of tools in educational contexts have clearly demonstrated the fact that individuals who use technological tools produce *instrumented schemes* that significantly influence the manner in which a target knowledge is constructed (see, for example, Béguin and Rabardel 2000; Vérillon and Rabardel 1995). Schemes refer to invariant patterns of action that individuals construct as a result of performing a task. The psychological tale speaks of schemes as mental structures that individuals develop from tasks independent of the tools that make them possible. The anthropological tale takes a much wider perspective by considering how schemes and tools mutually determine each other—that is, whereas tools enable the construction of specific schemes, the tool-generated schemes engender a further elaboration of the tools (see fig. 8.2).

Instruments (also referred to as tools) are either natural or made by people (artifacts). They are either material or semiotic (symbols and actions) in form. Vygotsky was certainly correct when he insisted that instruments "replace and render [as] useless a considerable number of natural processes, the work of which is developed by the instruments" (quoted in Vérillon 2000, p. 4). Further, he astutely pointed out how instruments are likely to "activate a whole series of new functions linked to the use and control of the instruments selected" (*ibid.*). Building on Vygotsky's (1978) insights and affirming them in his own work, Vérillon (2000) has articulated several cognitive processes that emerge from instruments that are "fundamentally different" from natural or "nonmediated" methods.

In this paper we shall, in particular, focus on mathematical processes that arise out of *instrumented activity*. Instrumented activity can be classified as either *pragmatic* or *epistemic* depending on the nature of the mathematical situation in which it is used. It is pragmatic when a tool is used in situations that alter an aspect of the physical world so that what emerges is practical, functional, technical, and situated knowledge. It is epistemic when a tool is used in situations that construct a specific kind of knowledge about the world with-

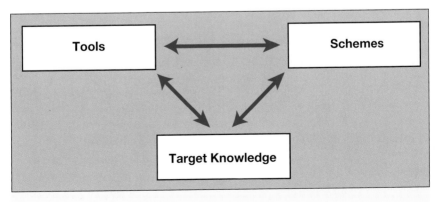

Fig. 8.2. Graphical model of instrumented schemes

out changing it, and such knowledge pertains only to those situations that exhibit regularities, that can be conveniently put into categories, and that are generalizable. The dynamic capability endowed in almost all technological tools that are currently available to learners (such as The Geometer's Sketchpad, Cabri, and most graphing calculators, such as TI-92, Casio FX, and Voyage 200) squarely fits within Vygotsky's (1978) perception about the *psychic* dimension inherent in any instrument. That is, these tools can exert a powerful influence on the actions of learners who use them. To use more updated terminologies, when learners in instrumented activity appropriate tools so that they become an integral part of their thinking, then their actions are manifestations of *instrumentation*. Further, when learners use the tools in situations other than what they have been originally intended to be used for, then the tool itself undergoes *instrumentalization* (Vérillon 2000). The term *instrumental genesis*, then, pertains to how individuals transform an instrument from its material or semiotic nature (i.e., as an artifact) to one that has been endowed with a psychological function as a consequence of instrumentation and instrumentalization (Béguin and Rabardel 2000; see fig. 8.3).

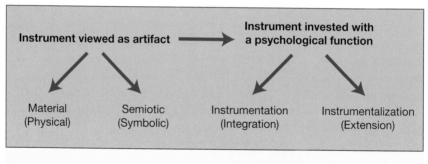

Fig. 8.3. Instrumental genesis

Geometric Tools and the Emergence of Proof and Related Instrumented Schemes

The three studies discussed in this section demonstrate the view that learners can develop proof skills in settings that are mediated by technological tools. Skeptics in years past have frequently pointed out the pernicious effects of calculators (such as the phenomenon of *deskilling*) on learners' mathematical processes without taking into account changes that were also needed if calculators replaced various time-consuming mathematical activities that involved procedural mastery. The studies furnish an account of how students develop instrumented schemes related to proof as a consequence of acquiring proficiency with the commands that have been built into the structure of Cabri, a dynamic software that can perform simple and difficult geometric constructions.[1] In particular, the ninth and tenth graders who participated in Mariotti's (2000) study developed notions of what it meant to think theoretically in mathematical settings with the aid of Cabri. In the study by Jones (2000), the twelve-year-old participants used Cabri to explore and establish relationships among different kinds of quadrilaterals. Further, their experiences with the geometric tool became the context they needed to develop deductive reasoning. The two pairs of high school students in Marrades and Guttiérrez's (2000) study furnish us with an exemplar of cases that model ways in which learners can use Cabri to discover for themselves what good proofs entail in school mathematics and how the dragging function, when used in an appropriate manner, could lead to different kinds of justifications.

Developing Theoretical Thinking in Technology-Mediated Classrooms

Mariotti was basically interested in determining how an interactive geometry software could influence the development of theoretical thinking among beginning high school students. Since the different commands in Cabri reflect the definitions, postulates, and theorems of Euclidean geometry, then Cabri's mediating power lies in its built-in capability to externalize geometric relationships for learners in a virtual, dynamic representational mode. This means to say that individuals who construct geometric objects in Cabri are implicitly learning the logical system of Euclidean geometry. There is, Mariotti (2000) insists, "a [potential] correspondence between the world of Cabri constructions and the theoretical world of Euclidean Geometry" (p. 28). Further, the "status of verification" changes when informal or intuitive geometry at the elementary school level is replaced by the deductive approach of secondary school geometry. Because not all constructions are possible with Cabri, those that do not work could be used as occasions for learners to reflect on why and where they fail. It is at this crucial stage when learners see the necessity of developing theoretical thinking.

In one of several experiments conducted by Mariotti, all the participants were told to construct a segment and draw a square whose side has the same measure as the segment. They were then asked to describe and justify their process of construction. One pair of students drew a square and insisted on

---

1. Perhaps arguable in many aspects, Cabri and The Geometer's Sketchpad share many common features because both have been developed from the same axiomatic structure of Euclidean geometry. Readers who are unfamiliar with Cabri may want to think of their experiences with The Geometer's Sketchpad, which is used more widely in our classrooms.

measuring both angles and sides to verify that the drawn square was consistent with the attributes of a square. When the teacher dragged the square through one of its vertices, the entire class agreed that it no longer represented a square. A second pair performed the following steps: (1) construct segment s_1; (2) construct line l perpendicular to segment s_1; (3) draw a circle where line l contains a diameter of the circle, segment s_1 a radius of the circle, and segment s_1 and line l intersect to contain the center of the circle and thus producing two consecutive congruent segments, s_1 and s_2; (4) determine two intersection points, one between line l and the circle (call this point P) and another between line l and segment s_1 (not the center of the circle; call this point Q); (5) construct segment s_3 perpendicular to segment s_2 through point P; and (6) construct segment s_4 parallel and congruent to segment s_2.[2] Since all four segments have the same length without actually measuring them, a square is formed. From the transcript that accompanied the construction, what surfaced from the ongoing interaction between and among the teacher, the class, and Cabri was the shift "from action to explaining the motivation of the action" (Mariotti 2000, p. 40). Because the teacher intervened mainly by posing questions, the class went beyond the figures being drawn when they had to assess the viability of the process of construction being performed. Some students also used the "history command" in Cabri to trace the step-by-step procedure that allowed them to observe and reveal the hidden geometric properties that were used in constructing the square.

The simplicity of the experiments used in the study reveal important instrumented schemes. Learning new commands in Cabri for the participants also meant learning the steps necessary for theoretical justification (see fig. 8.4 for a sample menu).

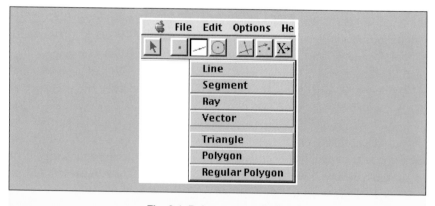

Fig. 8.4. Polygon menu in Cabri

---

2. Notations were introduced simply to facilitate the construction in words without having to draw the actual figures.

Developing a facility for manipulating the menus of the geometric tool in effect introduced the participants to the complexities involved in understanding the axiomatic structure of Euclidean geometry. When participants were asked to construct both the midpoint of a segment and a line perpendicular to another line, the procedures they identified reflected a level of internalization in which the external tools of Cabri have become "internal tools related to theoretical control [corresponding to using axioms, theorems, and definitions]" (Mariotti 2000, p. 43). This meant that they no longer treated geometric construction tasks as a mere procedural activity, because they have invested the construction of geometric figures with the necessary "theoretical meanings." Further, when participants got accustomed to the "externally oriented tool" of *dragging* in Cabri, all figures became dynamic (versus static). That is, dragging parts of the figures enabled them to explore multiple positions and possibilities and to formulate conjectures and prove or disprove them either by justifying them at both theoretical and visual levels or by developing counterexamples. (See Cabri Activity 1 in the accompanying CD for an illustration of ideas taken up in this section.)

Accounting for the Progressive Mathematization of Students' Geometric Thinking with Technological Tools

Jones (2000) undertook the difficult task of finding ways in which geometric tools can be employed to help junior high school students successfully make the transition from informal to formal methods of proof. He was especially interested in understanding how individuals "learn to reason" mathematically at the appropriate level, including how they develop explanations about geometric objects and relations in the dynamical environment of Cabri (pp. 57–58). Jones makes a distinction between verification and proof, whereby the former refers to a "demonstration of truth by some fact or circumstance" (p. 57) and the latter is a form of explanation that works within established rules that appear acceptable to a community of mathematicians. In his study, Jones carefully documented the mediating capability of Cabri as his participants attempted to develop a formal understanding of the different kinds of quadrilaterals. In so doing, he was able to write about the evolution of their "progressive mathematization"—that is, from their *perceptual* ability to recognize, classify, and describe different shapes of quadrilaterals to a more sophisticated *conceptual* ability of being able to define and deduce attributes of, and relations among, quadrilaterals.

Each participant in Jones's study was assessed to be of above-average mental ability. All of them studied in a progressive school where problem solving was strongly emphasized in all curricula. The study was divided into three phases. Phase 1 was meant to familiarize the participants with the features of the Cabri. Beyond that, they were asked to construct an exact, faithful copy on the Cabri screen of figures from paper in such a way that even if they were dragged, the transformed figures would still preserve the invariant characteristics. In phase 2, they were asked to construct specifically a rhombus, a square, and a kite that remained invariant even when dragged. They were also required to explain what made each figure a particular quadrilateral. At this stage, they were slowly being encouraged to construct for themselves their knowledge of the invariant properties of each quadrilateral. In phase 3, they investigated relationships between a rhombus and a square, a rectangle and a square, a kite and a rhombus, and a parallelogram and a trapezium, and the relationships among a rhombus, a rectangle, and a parallelogram. They were also asked to explain inclusive relationships, such as why all squares are rectangles or why all rectangles are parallelograms. The final task involved constructing a hierarchical model that would demonstrate the critical features of each quadrilateral and display relationships that exist among any two or more quadrilaterals. In exploring quadrilaterals with the Cabri, they were allowed to use all available commands except for the measurement tools, since Jones wanted them to establish geometric properties of quadrilaterals conceptually far beyond a mere numerical verification of dimensions, angles, and parts.

Jones identified the following stages of mathematization from the participants: (1) an initial ability to describe (versus explain) and reason perceptually (versus reason conceptually); (2) the emergence of situated forms of mathematical explanation that relies significantly on Cabri functions; and (3) an ability to provide mathematical explanations unaided by the geometric tool. Cabri's role is seen as central in the shift from the first to the third stage. Specific to the programmed conditions of dynamical tools, it is worth noting that the participants' exposure to the "dependency" mechanisms underlying the construction of figures in Cabri enabled them to reflect on the order in which all geometric objects are actually formed. For instance, two intersecting lines that meet at a point could not be dragged about the point because the point depends on the lines (i.e., the lines made it possible for the point to exist). The lines, however, could be dragged. Beyond this technological constraint, Cabri at stage 2 encouraged the participants to argue using *pragmatic explanations*—that is, explanations in which "the abstraction is 'situated' in that the knowledge is defined by the actions within a context" (Jones 2000, p. 81). An example of a pragmatic explanation is as follows: "[Y]ou can make a rectangle into a square by dragging one side shorter and so the others become longer until the sides become equal" (*ibid.*). Learners who progressed to stage 3 provided arguments using *conceptual explanations* that do not rely on situated abstraction or action. (Try Cabri Activity 2 on the accompanying CD to gain a better understanding of how learners proceed through the different stages.)

Exploring Different Types of Proof in Technology-Mediated Classrooms

The participants in Marrades and Gutiérrez's (2000) study explored thirty activities. Each activity was organized around three phases: constructing or exploring the features of a figure, verifying the correctness of conjectures, and justifying the conjectures they developed from an activity. For the participants, the dragging function in Cabri had assumed an essential role in assessing the validity of a construction. They knew that if transforming a figure by dragging would result in preserving many or all of its properties or relations, then the figure had passed the dragging test along those dimensions. For them, learning to apply the dragging function was seen as justifying claims at the empirical level. The use of follow-up questions such as "Why is the construction valid?" or "Why is the conjecture true?" meant having to provide justification that goes beyond the mere verification of facts or conjectures.

Marrades and Gutiérrez claim that interactive geometry software tools tend to change both the environment and the methods in which mathematical ideas are explored. Because learners are allowed to "experiment freely [and conveniently]" using "nontraditional" approaches provided by the tools, the manner in which justification and proof are employed will likely take many different forms. In their study, they demonstrated how some justifications were empirical, whereas others were deductive. Empirical justification, on the one hand, is characterized by an extensive use of Cabri drawings for both the demonstration and the justification of truths. Deductive justification, on the other hand, is far more decontextualized and does not rely on examples and their properties for proof. In fact, formal deduction involves both symbolic and abstract justification without having to refer to any example. However, examples could still be used in a deductive justification. When this happens, they are mainly employed for organizational purposes. Deductive justification in this situation uses a *thought experiment* that is of either the transformative or the structural type. *Transformative examples* are meant to verify the accompanying equivalent processes that take place in a deductive proof, whereas *structural examples* furnish step-by-step illustrations corresponding to the more formal steps used in the proof.

With empirical justification, when learners use examples to demonstrate the correctness of a conjecture, then this process is classified as *naïve empiricist*, that is, either of the *perceptual* (visual or tactile) or *inductive* (features or relationships) type. When the viability of a conjecture is established by using "a specific, carefully selected example," then this process uses a *crucial experiment*, and justification is *example-based* (since there are no other examples), *constructive* (how the example is obtained), *analytical* (using the known properties of the example), or *intellectual* (using the established properties of the example and, possibly, including other relevant properties that have been acquired previously). When the viability of a conjecture is established by using a *generic example*, then justification not only includes empirically verifying that the conjecture holds for the example (as with crucial experiment) but also includes referring to the critical features of the class of examples represented by the generic example. (Cabri Activity 3 on the accompanying CD involves a geometry problem that requires learners to provide both empirical verification and theoretical justification.)

Conclusion: Technological Ruminations on Mathematics Teaching and Learning

> It may be that we cannot see the truly new forms of rhetoric and theory that are emerging. What we see as senseless beauty may be the emergence of as yet unrecognizable new ways of making sense. [3]
>
> —Michael Joyce,
> *Of Two Minds: Hypertext Pedagogy and Poetics*

A significant number of mathematics teachers today perceive technological tools (graphing calculators, software with dynamic features, computers, computer-aided instructional resources, etc.) in the grades K–12 mathematics curriculum as mere add-ons to the plethora of already existing traditional tools (textbooks, paper, pencils, etc). The use of traditional tools has remained durable because they have proved to be necessary and useful in supporting cognitive functioning and processes. Learners use such prosthetic enablers as paper and pencil to perform complex operations and mathematical problems externally, and textbooks provide learners with both an organized presentation and an accessible and concrete representation of knowledge. The gradual acceptance of simple calculators in the grades K–12 classrooms was premised on the view that they were to be used mainly to augment what learners could not easily accomplish mentally. Teachers' resistance to—and desires to dismiss in the case of others—technological tools and mathematical learning in technology-based environments may be grounded on a naïve view that sees

---

3. This quote pertains to Joyce's pronouncement about hypertexts. But the same argument could be made about technological tools, given the agnostic reception accorded to them throughout the history of their evolution.

these new instruments of intervention as changing the nature and purpose of tasks that learners perform and nothing more. Although the tools and the new environment offer powerful motivation for learners, many teachers firmly believe in the psychological view that it is ultimately the learner who exerts the greatest influence in the construction of her or his mathematical knowledge. These two views partly explain why the history of technological tools in mathematics classrooms is rife with ambiguities. Skeptical teachers consider these innovations as being foisted upon learners' experiences unnecessarily.

Grounded in an anthropological context, technological tools furnish a more dynamic form of mediation that enables learners to develop mathematical knowledge. Viewed as a form of instrumented activity, mathematical learning in technology-based environments engages individuals and encourages them to construct and to experience what it means to produce knowledge. Tools start out as external devices for learners. As the tools undergo instrumentation, learners internalize them. Learners then develop instrumented schemes, and the process goes on again. In figure 8.5, the graphical model illustrates the bidirectional relations among tools, schemes, and target knowledge and could be interpreted in many ways. When learners use tools to acquire knowledge without developing meaningful schemes, then the knowledge they produce may be superficial, and the operations they perform tend to be procedural. Such knowledge is referred to as mere appearance matches. Individuals who develop schemes that eventually lead to knowledge is the typical story we obtain from different psychological theories of learning. However, it is almost impossible to conceive of schemes in the absence of any tool. The dynamical environment offered by new and emerging software and graphing calculators, for instance, allows for the emergence of different kinds and levels of mathematical processes involving pattern recognition, verification, explanation, jus-

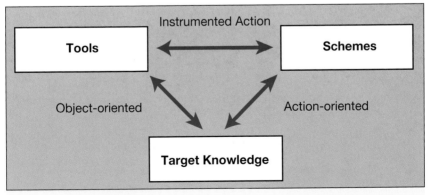

Fig. 8.5. Anthropopsychological account of mathematical learning in technology-based environments

tification, and proof. In an interesting short research paper about dragging in Cabri, Arzarello et al. (1998) wrote about instrumented actions that students in their study exhibited. Some used *wandering dragging* to explore and investigate possible features, attributes, and relations, and others employed *dragging test* to test for possible invariant properties. A number of them used *lieu muet dragging* to generate examples that satisfy a given property, and a few performed *bounded dragging* where a part of a figure is dragged and is investigated along a different objective primarily to determine the extent to which a feature of the figure is affected. (See Cabri Activities 4 through 6 on the accompanying CD for an illustration of each of the actions above.)

The studies discussed in this paper powerfully demonstrate a shift in thinking about the role of technological tools in the formation of both schemes and knowledge, that is, from tools as providing augmentation to tools as providing mediation. Tools used in mathematics not only make learners more capable. Tools also change learners in ways that affect the manner in which they construct and process knowledge. Tools mediate inasmuch as they are integrated in the internal and external activities and cognitive functioning of all learners.

References

Arzarello, Ferdinando, Chiara Micheletti, Federica Olivero, and Ornella Robutti. "Dragging in Cabri and Modalities of Transition from Conjectures to Proofs in Geometry." In *Proceedings of the 22nd Conference of the International Group for the Psychology of Mathematics Education,* vol. 2, edited by Alwyn Olivier and Karen Newstead, pp. 32–39. Stellenbosch, South Africa: Stellenbosch University, 1998.

Béguin, Pierre, and Pierre Rabardel. "Designing for Instrument-Mediated Activity." *Scandinavian Journal of Information Systems* 12 (2000): 173–90.

Jones, Keith. "Providing a Foundation for Deductive Reasoning: Students' Interpretations When Using Dynamic Geometry Software and Their Evolving Mathematical Explanations." *Educational Studies in Mathematics* 44 (2000): 55–85.

Mariotti, Maria Alessandra. "Introduction to Proof: The Mediation of a Dynamic Software Environment." *Educational Studies in Mathematics* 44 (2000): 25–53.

Marrades, Ramón, and Ángel Gutiérrez. "Proofs Produced by Secondary School Students Learning Geometry in a Dynamic Computer Environment." *Educational Studies in Mathematics* 44 (2000): 87–125.

Vérillon, Pierre. "Revisiting Piaget and Vygotsky: In Search of a Learning Model for Technology Education." *Journal of Technology Studies* 24 (2000): 1–12.

Vérillon, Pierre, and Pierre Rabardel. "Cognition and Artifacts: A Contribution to the Study of Thought in Relation to Instrumented Activity." *European Journal of Psychology of Education* 10 (1995): 77–101.

Vygotsky, Lev. *Mind in Society.* Cambridge, Mass.: Harvard University Press, 1978.

PART 2

NOTES FROM THE FIELD

9

The Reality of Using Technology in the Classroom

Suzanne Alejandre

REMEMBER this scenario? You attend a session at a conference presented by two teachers who talk about how they have used technology to teach mathematics. They give a demonstration of a Java applet[1] they use with their students. They even show a video of the students working on computers in a lab. You can see how the students are engaged in mathematical thinking, and you think, "I am going to start using technology in my classroom!"

You enter your classroom on Monday and reality hits. There are so many things to consider before you start using technology. You have the information from the session, but the computers described in the handout are different from what your school has. You do not have the same software. You are not sure how to sign up for the lab or even if mathematics teachers have access. You have a computer in your room, but you do not have one of those fancy video projectors.[2] You think maybe the school has one, but you are not sure if you can check it out, and even if you did, how would you make it

---

1. Java applet: A short program written in a computer language called Java. It is viewed and used interactively on the Internet using a Web browser.

2. Video projector: A device that can be connected to a computer to project a large-screen display.

Editor's note: The CD accompanying this yearbook contains an electronic version of this article, complete with hyperlinked URLs.

work? You start worrying that attending to all of these details will take too much time, particularly because you do not have enough time to do everything already! Suddenly you are back where you started. You have decided that it is just too hard to try using the new ideas that sounded great before you started considering the details. Maybe there is a way, however. With some realistic thinking, there are answers to all the doubts and questions. Perhaps the scenario that you were shown at the conference cannot be copied, but that does not mean that you cannot begin incorporating the use of technology into your classroom instruction.

How can a teacher start? Slowly, is an important part of the answer; also, there is the need to think realistically. Assess what tools you have available and begin with one idea. As you incorporate technology into your teaching, build slowly. Find a way to talk with other teachers who are using technology to get ideas, but realize that each teacher may have a different setup. It is important to adapt other settings to the setting that you have. Instead of using the excuse that you do not have what they have, consider what you do have and go from there! Concentrate on what you can do with what you have got. As you use what you have, perhaps with time you will find ways to add more capabilities, but at the beginning that should not stop you from starting.

Following are some of the realistic situations that classroom teachers face along with suggestions on how to make the best of them.

Equipment

• I have only one computer in my classroom. It has no Internet access.

Use the computer as a "reference station." There are often occasions during a mathematics lesson when having a fact checked by one person can add to the task that the class is completing. Reference software possibilities include calculator programs or a mathematics reference CD ROM, such as an almanac or glossary. Assign in turn a student, a pair of students, or even a group of four students to work on the computer and report back to the class.

Calculator example
Many models of computers come with a calculator utility already loaded on the hard drive. It's also possible to download a calculator program if you have access to an Internet-accessible computer. You can copy it to a disk or CD, or you can download it directly to your classroom computer.

Tucows—Windows freeware/shareware calculator downloads
www.tucows.com/calculators95_default.html

Tucows—Macintosh freeware/shareware calculator downloads
allmacintosh.xs4all.nl/interface_calculators_default.html

Glossary or almanac example
When CD ROM drives were sold separately from computers, they often came packaged with a variety of CD ROMs. Check your department, school, or district library. They may have CD ROMs available for checkout.

Use the computer to accompany and enhance a mathematical data lesson. If you are collecting data as part of your lesson, have a student, or a pair of students, enter the data into a spreadsheet as the rest of the class are collecting the data. This activity will accomplish two things: (1) it will save time, and (2) it can result in a display of the data that can be manipulated to demonstrate the mathematics you are discussing.

Spreadsheet example
Leonardo da Vinci activity
mathforum.org/alejandre/frisbie/math/leonardo.html

Students are asked the question: Is the ratio of our arm span to our height really equal to 1? Data are gathered to determine how the students should respond.

Use the computer as one of the stations[3] within a menu of stations. Stations or math centers can be used to introduce concepts or reinforce concepts. Arrange the classroom so that there are areas for groups of three or four students to work on an activity. For example, if you have thirty-six desks, arrange them so that they are facing one another in nine groups of four. It's always good to have one or two more areas than there are groups of students so that students are able to move around as they complete an activity. A countertop or side table can be used for the additional stations.

Classroom "menu of stations" example
If probability is the topic, ten different probability activities could be set up in different areas of the classroom. One of the stations could

---

3. Stations (learning stations or learning centers): An instructional technique often used at the elementary and middle school level where separate areas are set up in the classroom. Small groups of students rotate through the centers and work on the tasks using the instructions given at the station. Usually the stations each have a common theme so that when the center's activity has been completed, the class can debrief and come to some consensus about the mathematics that was learned.

take advantage of your single computer if it were loaded with some probability software like the following two examples:

Statistics Education Research Group— Prob Sim
www.umass.edu/srri/serg/software.html

Probability Explorer
www.probexplorer.com/

Schoolwide "menu of stations" example
Multicultural Math Fair
mathforum.org/alejandre/mathfair/index.html

This site gives ten activities complete with set-up directions and worksheets in English and Spanish. It offers many software applications in downloadable form: Hypercard, The World Stack, and HyperStudio are a few choices.

Another model of having a selection of stations is to create them so that there are only four different activities, all clustering around the same concept. The variety lies in the tools that are used to reach the concept. One of the activities has a "computer tech" focus; another has a "calculator tech" focus; a third has a manipulative focus; and the final activity has a paper-and-pencil focus.

Computer example
Building Perspective by Sunburst Technology
www.sunburst-store.com/cgi-bin/sunburst.storefront/EN/
Product/9370

Building Perspective is a game designed to help students in grades 4 and up develop their spatial perception and visual reasoning skills. The object is to figure out how a 3 × 3, 4 × 4, or 5 × 5 city will appear when viewed from above. Because the buildings range from one to nine stories tall, some may be completely hidden by others in some of the views. Students need to use their spatial sense, their ability to collect and organize data, and their information-gathering ability to place all the buildings.

Graphing calculator example

Building Perspective—TI Graphing Calculator application
education.ti.com/us/product/apps/buildperspec.html

An equivalent application is available from Texas Instruments to be used with the TI-73 graphing calculator.

Manipulatives example

Use Unifix cubes, so that the buildings can be constructed to correspond to the software activity.

Isometric dot paper example

mathforum.org/alejandre/escot/isometric.dot.gif

Use the isometric dot paper to develop a paper-and-pencil activity that corresponds to the software activity.

• I have only one computer in my classroom, but it has Internet access.

Use the computer as a "reference station" using any of the many interactive Web sites.

Reference site examples
The MacTutor History of Mathematics
www-history.mcs.st-and.ac.uk/

This is the premier site on the Web for math history. Major features include indexes for biographies, history topics, famous curves, mathematicians of the day, and a site search engine.

ConvertIt.com Measurement Converter
www.convertit.com/Go/ConvertIt/

ConvertIt is Entisoft's comprehensive converter for distance, area, volume, mass, speed, temperature, pressure, energy, power, force, and so on—more than 500 different unit conversions.

Ask Dr. Math Frequently Asked Questions
mathforum.org/dr.math/faq/

Frequently asked questions from the archives of the Math Forum's question-and-answer service for grades K–12 math students and their teachers, with classic problems, formulas, and other recommended mathematical sites.

Use the computer as suggested in the "no Internet access" section but without the need to transfer the programs that are downloaded.

Sites that enhance the mathematics lesson examples
Industrious Clock by Yugo Nakamura
www.yugop.com/ver3/stuff/03/fla.html

This Flash animation is a short demonstration of elapsed time.

Create A Graph
nces.ed.gov/nceskids/Graphing/

Create A Graph is an interactive tool for creating area graphs, bar graphs, horizontal bar graphs, line graphs, and pie charts. After selecting one of the types of graphs, there is an explanation to help the user decide if this is an appropriate graph for the given data.

Math Tools
mathforum.org/mathtools/

Math Tools is a community and library dedicated to cataloging interactive tools to enhance mathematics teaching and learning.

• I have one computer and a display (video projector or television monitor).

With the addition of a method to display the computer screen to students, one computer in the classroom can become much more versatile!

Use the computer to introduce a new concept. If you do not have Internet access, you can display text, drawings, diagrams, or problem prompts that previously you might have displayed using an overhead or chalkboard. If you have Internet access, display an interactive page (Java, Flash,[4] Shockwave,[5] etc.) to prompt conversation about the topic.

Internet-accessible examples
Student I-Math Investigations Grades 9–12: Investigating Linear Relationships: The Regression Line and Correlation

illuminations.nctm.org/imath/912/LinearRelationships/student/
index.html

---

4. Flash: Macromedia, a technology company, developed this type of program similar to a Java applet. It is viewed and interactively used on the Internet by means of a Web browser.

5. Shockwave: Shockwave, developed by Macromedia, is a family of multimedia players. Users with Windows and Mac platforms can download the Shockwave players from the Macromedia site and use it to display and hear Shockwave files. Shockwave is especially popular for interactive games.

6. Interactive software: When a program or screen is described as interactive, users can use the computer mouse to move items on the screen and enter numbers, and the program responds to the input.

High school students investigate the relationship between a set of data points and a curve fitted to those data points by least squares using an interactive[6] linear regression tool.

Varnelle Moore's Primary Math Activities: Equal Parts
mathforum.org/varnelle/knum1.html

Elementary school students learn to divide a circle into pieces of equal size.

Project Interactivate: Slope Slider

www.shodor.org/interactivate/activities/slopeslider/index.html

This activity allows for the manipulation of a linear function of the form $f(x) = mx + b$ and encourages the user to explore the relationship between slope and intercept in the Cartesian coordinate system.

Use the computer to develop mathematical understanding of graphical data representation. Use a spreadsheet program such as Claris Works or Excel or use one of the Java applets available on the Internet.

Leonardo da Vinci Activity
mathforum.org/alejandre/frisbie/math/leonardo.html

After the data have been gathered and entered into the spreadsheet, display the graphs that can be generated from the spreadsheet data and discuss what the graph shows.

Project Interactivate: Bar Graph
www.shodor.org/interactivate/activities/bargraph/index.html

The Bar Graph applet allows the user to graphically display data frequency using a bar graph. The user is able to enter his or her own data as well as manipulate the y-axis values. The ability to manipulate the y-axis values allows for the creation of potentially misleading graphs.

Use the computer to review concepts either by using a word-processing program to type questions into a document or by using Internet-accessible activities.

Mrs. Glosser's Math Goodies: Operations with Integers
www.mathgoodies.com/lessons/vol5/operations.html

There are many rules to remember when learning to add, subtract, multiply, and divide integers, and it is easy to get confused. This lesson reviews all the rules learned for operations with integers. It includes an online, interactive mixed set of exercises to complete.

Functions and Their Graphs: Quiz
math.dartmouth.edu/~klbooksite/1.03/103quiz/103quiz_index.htm

Quiz is a calculus course quiz (with answers) on functions, range and domain, and odd and even functions.

Use the computer to play basic-skills games. Divide the class into two groups. Have one student from each group use the computer, taking feedback from their teams.

Examples
Project Interactivate: Fraction Four
www.shodor.org/interactivate/activities/fgame/index.html

Fraction Four is a game to practice multiplying, dividing, comparing, and converting fractions, decimals, and percents.

Project Interactivate: Maze Game
www.shodor.org/interactivate/activities/coords/index.html

The Maze Game allows students to have fun while mastering their understanding of how the Cartesian coordinate system works by maneuvering a robot through a minefield.

Use the computer to introduce what you will have students do when they go to the computer lab. Briefly presenting your computer lab expectations before taking students to a lab can help things run smoother.

• I have one cluster of four computers.

Depending on how many students you have in a class, these computers can be used in different ways. If you have thirty-six students in a class, then it limits you. But there are still options!

Use the four computers as a reference area. If you had eight groups of students, one student from each group could go to the computers, which would mean that a pair of students would be at each computer. Those students' job would be to get the assigned information and report back to their group. Similarly, this could be done as an assignment for the whole class. In that situation, you could have four or eight students assigned as computer "monitors" who would report back to the class. If you rotated this assignment, all the students would have their turn in that role.

Example

> Using the Ask Dr. Math Archives
> mathforum.org/workshops/sum2000/link/

> This site contains a list of link pages compiled from the Math Forum's Ask Dr. Math service to use with students to find information on a particular mathematical topic.

If you are using the menu of stations idea that was explained previously, you could have four technology-based stations set up in your classroom. You could have four different activities, or you could have the same activity at each station with fewer rotations of student groups in order for each group to have an opportunity to work with the technology.

Example

> Set up the four computers with these four different virtual manipulatives that are available online or on a CD ROM. The nontech stations could have corresponding paper-and-pencil activities to reinforce students' understanding of visualizing, naming, comparing, and writing equivalent fractions.

>> National Library of Virtual Manipulatives: Visualizing
>> matti.usu.edu/nlvm/nav/frames_asid_103_g_2_t_1.html

> Students can divide a shape and highlight the appropriate parts to represent a fraction.

>> National Library of Virtual Manipulatives: Naming Fractions
>> matti.usu.edu/nlvm/nav/frames_asid_104_g_2_t_1.html

> Students write the fraction corresponding to the highlighted portion of a shape.

>> National Library of Virtual Manipulatives: Comparing Fractions
>> matti.usu.edu/nlvm/nav/frames_asid_159_g_2_t_1.html

> Students judge the size of fractions and plot them on a number line.

>> National Library of Virtual Manipulatives: Equivalent Fractions
>> matti.usu.edu/nlvm/nav/frames_asid_105_g_2_t_1.html

> Students manipulate the applet to understand the relationships among equivalent fractions.

- ## I have no computers at school, but my students all have home accessibility.

Assign homework and projects that involve technology. Encourage students to use spreadsheets, word processing, graphs, Web references, and so forth, when appropriate. If you have been modeling good uses of technology, encouraging students to use it as they complete their homework and projects is an opportunity to have students follow that model.

Examples of sites students might access:

Math Tools
mathforum.org/mathtools/

Ask Dr. Math
mathforum.org/dr.math/

Problems of the Week
mathforum.org/pow/

- ## I have to sign up to use a computer lab.

Following are a few tips that might help this situation go smoothly.

Create a numbering system for the computers in the lab. Either physically or mentally number the computers. Before bringing students to the lab, assign them a number and explain how the numbering scheme works in the lab. This system helps maintain accountability if students misuse the equipment. It also helps you have students sit where they can help one another.

Use "help" signs (some labs use paper cups) in the lab. If a student has a question, tell them to put the "help" sign on the top of their computer monitor. If many help signs suddenly appear, you will know that it is time to interrupt the students' work and address the class. If just a few help signs appear, you can easily identify who needs help and move around the room and assist them.

Have a backup plan. If unforeseen technology glitches occur, have something ready for the students to work on. Perhaps a student could lead an oral math game while you are trying to fix the problem. Or you could have the students bring their textbooks so they can read a certain section if you are preoccupied.

- ## I have no way for students to save their work.

Probably you just have not thought of a reasonable system to use, but consider a few of the following possibilities. Not all work needs to be saved, but if you would like this feature, it is a good idea to have a plan figured out.

Are your computers of the "disk" era? If they are, probably sets of data disks are in a drawer or cupboard at your school. Devise a system that will prevent students from deleting one another's files, and so on. The best idea is to have one disk for each student. If you have several classes, keep each class set separate and clearly labeled. Consider adding a paragraph about "use of data disks" to your rules for computer use.

Are your computers of the "no disk" era? Consider creating a system of directories on your computer(s). Taking the time to organize the system and make it clear how it should be used might save you problems in the future.

Are your computers connected to a server? It is very possible that the school site has a system of directories already organized for student work. Ask questions. Find out what policies your school has in place.

Age of Equipment

• Our computers are too old.

The best defense for this "problem" is to accept the capabilities of the equipment that you have and make the best of it. A good analogy is to think of how you treat an old car. You may not be able to drive it as fast or rely on it as much as you would like, but it still has the capability of getting you where you want to go. Right? One use of older computers is the almost universal ability to run basic spreadsheet capabilities. Also, the more basic the computer, the more likely the student will focus on the task at hand. Computers that seem outdated still have a lot of possibilities. Try to maximize them!

• We have the old models of graphing calculators.

Companies that produce graphing calculators have continually been improving them and coming out with new models. Look for user groups on the Web. Plenty of people still use the "old" models. Remember that even if the calculator is an old model, it still produces correct mathematics.

Software

• I have only one copy of the software I'd like to use.

You have two choices. One possibility is to make use of the one copy you have by using it as described above under the "one computer" category. Another possibility is to find a way to buy a site license! Sometimes you can convince your school to purchase a site license for software if you can demonstrate that it is a good tool. The way to do that is to use the one copy that you have effectively. One idea might be to take photographs of eager students waiting to use the one computer and show them to teachers, administrators, and parents as a tool to lobby for additional computer purchases.

• I have an "old" version of the software that I saw demonstrated.

It can be frustrating not to have exactly the same materials that attracted you to the idea, but think of the educational objective that you are trying to meet.

• We don't have any of the software that I saw demonstrated.

Think about the functionality of the software. How did it help students learn mathematics? Once you have the functionality in mind, look through the software that you have available. Is any of it comparable? If nothing comes close, consider checking for sources either at your school site or in the school district. Sometimes libraries of software are available that can be checked out, or your site might even have a site license for the particular software that you were unaware of. Another possibility is that you can download a demo version of the software. It may not have all the features or it may be good only for a limited time, but it would give you a chance to see if it is worth purchasing the full version.

Internet Accessibility

• I'm afraid that students will go to inappropriate sites.

Many school districts have "acceptable use" policies that students, teachers, and perhaps even parents or guardians are required to sign before students use the Internet. Look into the guidelines that your school uses and follow them.

It's possible that your school district has site-blocking software that filters access to inappropriate sites. Check with the technology department at either your site or district. This may alleviate your fears.

Regardless of what measures are in place, be mindful of what actions need to be taken if a student goes to an inappropriate site. Treat it as you would any other behavior problem. Act decisively.

The best defense against students going to inappropriate sites, however, is to have specific, interesting, and mathematically useful tasks for the students to do.

- **We have Internet accessibility but old computers, and there are so many problems.**

Out-of-date computers can definitely slow things down. Sometimes adding more memory to an older computer can help. The most dramatic solution is to forget using those computers with the Internet and try some of the non-Internet uses of the computers. Again, the best defense is to accept the computers' capabilities and maximize their use but understand their limitations. Accept them for what they can do well and use them for those purposes.

- **It seems that the sites that look good always need plug-ins,[7] but all the computers are "locked."**

A little planning can be helpful. If you have found a good site, perhaps through using your home computer, make sure to check the site on your school's computer before assuming that it will work.

Do not assume, however, that you have to do all the checking. Do you have a student who might be able to preview sites that you would like to use with a class? Is a technology teacher or assistant available who could help?

Accountability

- **I already have lesson plans that address the standards that I'm required to teach. Why should I start using technology?**

Many districts and states have well-defined objectives that teachers are required to follow to enable their students to master a defined set of mathematics standards. If you have been using teaching strategies that are very successful at helping your students master certain concepts, it is probably better to leave well enough alone. Think instead of some concepts that your students continually struggle with and try using technology to help them with those concepts.

- **Technology seems to take so long; I just don't have the time.**

Start small. If you teach at the middle school or high school level, start with just one of your classes.

If you are prepared and understand the realities of using technology in the

---

[7] Plug-in applications are computer programs that can be downloaded, installed, and used as part of your Web browser. After they have been installed, Web pages that require that particular plug-in will function correctly.

classroom, its actual use will go more smoothly. Learn both the strengths and the weaknesses of what you have available. Technology in the classroom is just one more tool to help your students learn mathematics, and you want to make maximum use of all the tools that you have.

After you have acquired some preliminary experience using technology, you will know more details about what you have available for equipment, software, and the capabilities of both. You will be able to make more informed decisions, and you will continually be able to build on your experiences. As you attend the next technology presentation or conference, you will be armed with the knowledge of what you have, what you have experience using, and what you want to try in the future. You will be on your way to successfully integrating the use of technology as a tool for mathematical understanding!

10

Using Technology to Foster Students' Mathematical Understandings and Intuitions

Eric J. Knuth
Christopher E. Hartmann

TECHNOLOGY offers a unique and powerful means of fostering students' understandings and intuitions of the mathematics they study and, accordingly, should play an important role in classroom instruction. For example, interactive geometry technologies (e.g., Geometer's Sketchpad, Cabri) provide the capability to create models that allow classroom participants to engage in conceptual conversations about mathematical ideas and problem-solving methods. By a "conceptual conversation," we mean a conversation that has a diminished emphasis on technique and procedures and an increased emphasis on relationships, images, and explanations (Thompson 1996). In this article, several illustrations of how technology might be used to engage students in such conversations and, consequently, to foster students' understandings and intuitions are presented and discussed. Although the illustrations we present focus on a particular technology, they are meant to be demonstrative of ways in which technology more generally can be used to promote students' engagement with, and learning of, mathematics. Consequently, we encourage readers to reflect on the ideas we present and consider how they might use technologies in their own instructional practices to engage students in conceptual conversations.

Viewing the Relationships in a System of Equations

Systems of equations are a regular topic in most (if not all) algebra courses. Students typically learn to solve a system of equations using several approaches, including a graphical approach. Acquiring an understanding of a graphical solution to a linear system requires a student to understand both the meaning of

---

Editor's note: The CD accompanying this yearbook contains three Geometer's Sketchpad files and a listing of the hyperlinked URLs for the Web sites that are relevant to this article. For more information on The Geometer's Sketchpad, contact Key Curriculum Press at www.keypress.com/.

the graphical representation (i.e., that a graph represents the coordination of two quantities) and the models (graphical and algebraic) that describe the system. In the example that follows, several ideas necessary for understanding a graphical approach for determining the solution to a system of equations are developed.[1] The instructional progression that unfolds illustrates one way to foster students' insight into, and understanding of, the graphical solution process.[2] The following problem serves as the context for the illustration:

> Joshua is trying to decide which of two phone plans would be the better deal. Plan A has a monthly charge of $6.50 and costs $0.25 a minute; Plan B has a monthly charge of $1.25 and costs $0.60 a minute. Which plan is the better deal?

In solving this problem with a graphical approach, students are typically instructed to graph each plan and then to identify the intersection point—the point at which the plans cost the same amount at a particular point in time. Next, decisions are made regarding which plan is the better deal (e.g., Plan A is less expensive than Plan B after a specified amount of time). Yet, there are several important ideas underlying this approach that may not be discussed (because we often assume that students already have the prerequisite understandings). Technology offers a means of highlighting these ideas rather easily; as a result, the ideas can become a focus for a conceptual conversation (see the corresponding Geometer's Sketchpad [GSP] sketch on the CD). For example, figure 10.1 provides an opportunity to highlight the connection between the y-intercept of the graph for Plan A and the relevant information from the problem context—the monthly charge of $6.50. A second important idea, the dependence of the cost of the phone plan on the amount of time one has talked on the phone (i.e., "elapsed time") is easily demonstrated dynamically by dragging the point representing the elapsed time along the x-axis and observing the corresponding changes in the plan's cost (i.e., changes in the length of the line segment along the y-axis). For instance, as the time one has talked on the phone increases, figure 10.2 (in comparison to fig. 10.1) makes visually apparent the subsequent increase in the overall cost associated with Plan A (displayed as an increase in the length of the line segment along the y-axis).

An important power of the graphical representation of a function, in general, lies in the coordination of two distinct (covarying) quantities as a single

---

1. We invite readers to also review the e-example of the *Principles and Standards* that focuses on rate of change in linear functions—an idea central to understanding the systems of equations illustration we present (see standards.nctm.org/document/eexamples/chap6/6.2/index.htm).

2. We want to credit Patrick Thompson, a mathematics education professor at Vanderbilt University, for the idea of using interactive geometry software to explore a system of linear equations.

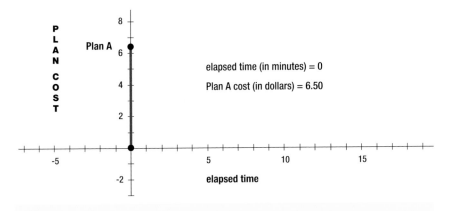

Fig. 10.1. Interpreting the y-intercept

point. In our example, adding a correspondence point (point *C* in fig. 10.3) gives us a means of simultaneously representing the covarying quantities of elapsed time and phone plan cost. Now, by dragging the point representing the elapsed time along the *x*-axis, students can see not only how the cost of Plan A changes (displayed as an increase in the length of the line segment along the *y*-axis) but also how the correspondence between the elapsed time and plan cost are represented graphically (point *C*). To further highlight this correspondence for students, a teacher might explicitly discuss the connection between the graphical representation and the corresponding symbolic representation—how, for example, is the cost for Plan A determined if given an

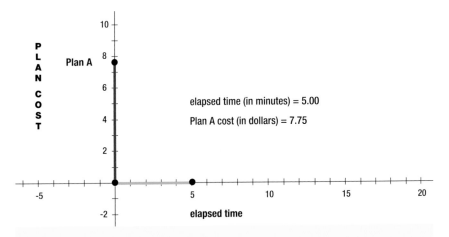

Fig. 10.2. The cost of Plan A increases as the elapsed time increases.

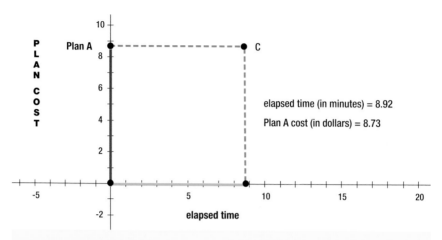

Fig. 10.3. Representing the correspondence between elapsed time and plan cost

elapsed time of 8.92 minutes (see fig. 10.3)? Similarly, a teacher might have students consider several values of elapsed time and associated plan costs as well as how these costs are determined (see fig. 10.4). At this point, students will likely have an idea of what the graphical representation of the phone plan might look like if one continuously recorded the correspondence between the plan cost and elapsed time as the elapsed time changed. One way of helping students to visualize the "final" graphical representation is to have them imagine that the correspondence point is sprinkled with a dustlike substance

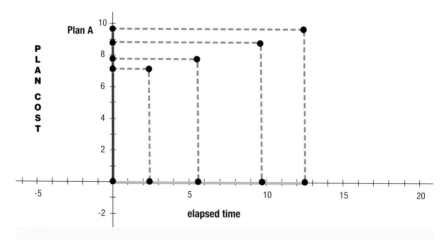

Fig. 10.4. Representing the correspondence for several elapsed time values

that leaves a trail behind as the point moves (e.g., using the Trace feature in Geometer's Sketchpad). Thus, increasing the elapsed time from 0 minutes produces the "dust trace" shown in figure 10.5.

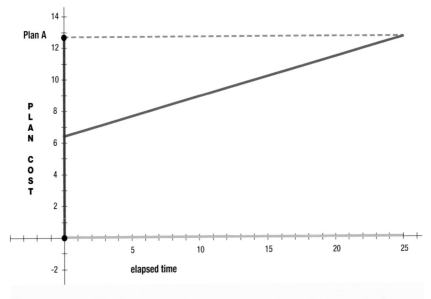

Fig. 10.5. Representing the correspondence as a continous function

After Plan A and its corresponding graphical representation have been discussed, a discussion of both Plan A and Plan B is appropriate. In this example, figure 10.6 displays the points that represent the correspondence between plan cost and a particular elapsed time for both plans. Again, discussing with students different features of the dynamic version of the graph displayed in figure 10.6 (e.g., what the y-intercepts represent, what the correspondence points represent, what the vertical distance between the correspondence points represents) will help to further strengthen their understandings not only of the graphical approach to solving a system of equations but also of graphical representations more generally. Finally, the traditional "product" of the graphical solution approach—the complete graph for each plan—can now be presented in a manner that builds on the understandings and intuitions students have developed as a result of the preceding presentation and discussion. Once again, sprinkling the correspondence points with the dustlike substance results in the graph displayed in figure 10.7. Students' decisions regarding

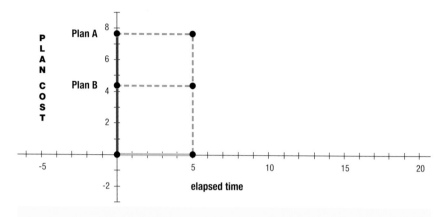

Fig. 10.6. Representing the correspondence for both phone plans

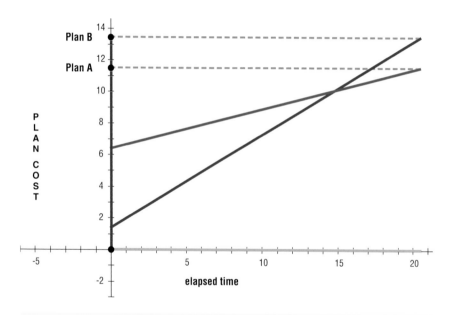

Fig. 10.7. Representing the correspondence for both phone plans as continuous functions

which is the best phone plan can now be grounded in a more conceptual manner than might typically occur.

The dynamic nature of this illustration allows a teacher to focus students' attention on aspects of the graphical solution approach that might otherwise be overlooked or difficult to highlight without technology. Moreover, this illustration enables teachers to complement a more traditional instructional strategy (i.e., producing the graph and observing the point of intersection) by engaging students in conversations about the conceptual issues related to both graphing and solving a system of equations using a graphical approach. Again, we encourage the reader to think about how other technologies might be used in fostering students' intuitions and understandings of systems of equations. For example, one might use a spreadsheet to display (in tabular form) different costs associated with each plan for varying amounts of time; representing the phone plan data in this way offers further opportunities for engaging students in conceptual conversations.

Exploring Images of Fit in a Best-Fit Line

As a second example of using technology to foster students' understandings and intuitions of a topic that is, again, a mainstay of secondary school mathematics, consider the topic of fitting a line to a set of data. A common instructional practice is to have students plot the data on a coordinate plane, and then ask them to use a piece of spaghetti to represent the line that they will "fit" to the data. Students are typically instructed to position the spaghetti noodle so that it appears to be as close as possible to each point—visually determining the "best" fit; such a placement for four data points might look like the line displayed in figure 10.8. At this point, students might determine the equation for

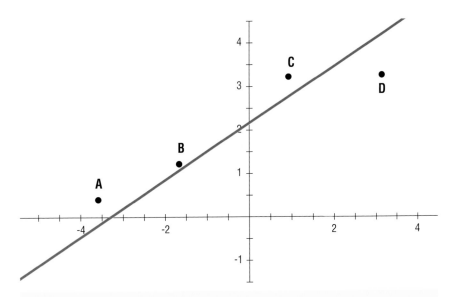

Fig. 10.8. A "visually" positioned line of best fit

their line and then use that equation in making predictions about additional points. Alternatively, the objective for the lesson might be to determine a line of best fit analytically, usually by using the statistical capabilities of a graphing calculator, and then to use the resulting equation in a similar fashion (i.e., to make predictions). In the former situation, the line that students identified as their line of best fit has not been determined mathematically and may or may not be the best fit in reality. In the latter example, the line has been determined mathematically, but students may not have an understanding of "what the calculator did" in determining the equation for the line or why the line is called a least squares line of best fit (the most commonly used line of best fit). Moreover, teachers often may not attempt to explain the underlying ideas, since the focus of the lesson may be on the use of the equation for the line. In either situation, ideas underlying the least squares line of best fit are not beyond the grasp of students and should be a topic of discussion.

Interactive geometry software offers a means of engaging students in a conceptual conversation about the least squares line of best fit (see the corresponding GSP sketch on the CD).[3] Figure 10.9 displays two lines with the images of their corresponding residual[4] squares visible. Presenting the topic of least squares line of best fit with this sketch provides a direct connection to why such a line is called a "least squares" line. The idea underlying a least squares line of best fit is to minimize the sum of the areas of the squares formed by the line segments connecting each of the actual or observed data points and their corresponding predicted values (i.e., the y-coordinates on the line associated with the x-coordinates of each observed data point). In observing the two lines displayed in figure 10.9, the first line is clearly not as good a fit as the second line, since the sum of the areas of the squares for the first line is 6.82 square inches and the sum of the areas of the squares for the second line is only 1.47 square inches. With this sketch, the dynamic features of Geometer's Sketchpad allow students easily to manipulate the slope and the y-intercept of the line as they seek to determine the line of best fit. By capitalizing on similar capabilities of Geometer's Sketchpad, such as the system of equations example, teachers find that the line-of-best-fit sketch allows them to use conceptual conversations to foster students' understandings and intuitions of fitting a least squares line to a set of data.

---

3. We recommend that readers also review the e-example in *Principles and Standards* that explores in further detail the concept of least squares regression (see standards.nctm.org/document/eexamples /chap7/7.4/index.htm).

4. A residual is the difference between the observed data point (e.g., the y-value of point A) and the predicted value (e.g., the y-value on the best-fit line).

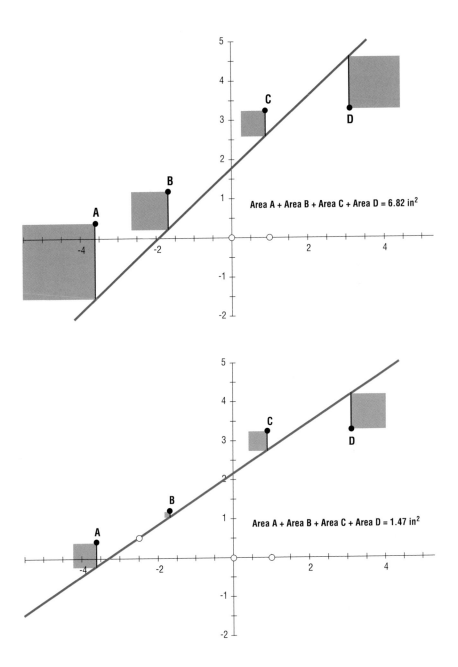

Area A + Area B + Area C + Area D = 6.82 in²

Area A + Area B + Area C + Area D = 1.47 in²

Fig. 10.9. Lines of best fit with the residual squares visible

Using Dynamically Linked Models to Generate Explanations

As a final example of how technology can furnish an opportunity for teachers to foster their students' understandings and intuitions, let's consider the graph of the sine function and its relationship to the unit circle. In this example, we address the following question: *Why does a graph of the sine function behave as it does?* Many teachers (sans technology) likely address this question by drawing a unit circle (including a central angle) accompanied by a graph of the sine function on the board and then using these drawings to highlight important connections between the representation of the sine function on the unit circle and its representation on the Cartesian plane for several "critical" angles (e.g., $\pi/2$, π, $3\pi/2$, 2π). Although this strategy can illustrate many important connections for students, the static nature of the presentation makes it somewhat limited. When the question above is addressed with the use of interactive geometry software (see the corresponding GSP sketch on the CD), however, one can explore the connections in a more dynamic and potentially more powerful manner (the Geometer's Sketchpad construction used in the illustration that follows can also be found in Bennett 1999).

Initially, teachers might foster students' intuitions by focusing students' attention on each of two particular unit circle quantities that covary—the length of arc *DA* and the height of point *A* above $\overline{ED}$ (see fig. 10.10)—and whose correspondence produces the graph of the sine function. Consider the first quantity, the arc subtended by the central angle *AED*—arc *DA*. As point *A* is dragged along the unit circle, the length of arc *DA* clearly changes as well. By dynamically linking the unit circle and a Cartesian graph, changes in

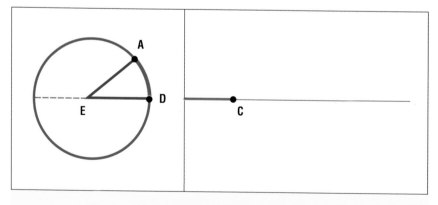

Fig. 10.10. Tracking the distance around the unit circle

the arc length due to the movement of point *A* along the circumference are reflected on the Cartesian graph by changes in the position of point *C* along the *x*-axis (i.e., the length of arc *DA* equals the distance between point *C* and the origin). Now let's consider the second quantity, the height of point A above *ED* —segment *AF* (see fig. 10.11). In this sketch, one can "track" changes in the height of point *A* (i.e., the length of segment *AF*) as the angle measure changes; in other words, one can observe changes in the value of the sine function as the measure of angle *AED* changes. As before, dynamically linking the unit circle and the Cartesian graph allows one to drag point A along the circumference of the unit circle and see the corresponding changes in the value of the sine function reflected along the y-axis (i.e., the location of point *B*). At this point, it is natural to discuss with students the value of the sine function for different angle measures (e.g., $\pi/2$, π).

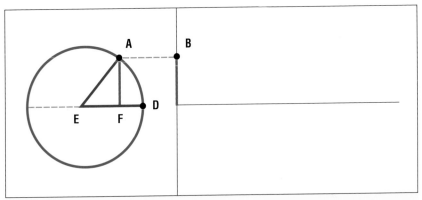

Fig. 10.11. Tracking the "height" of the central angle

In a fashion similar to that illustrated earlier with the phone plan example, we now coordinate the correspondence between the two quantities as a single point, *F*, in the Cartesian plane (see fig. 10.12). Now as one drags point *A* along the circle, the positions of both points *B* and *C* change accordingly, and thus we have the position of the correspondence point *F*. Finally, to visualize the correspondence between the two quantities as a continuous function, we again imagine that the correspondence point is sprinkled with a dustlike substance that leaves a trail behind as point *F* moves (see fig. 10.13). The dynamic link between the two models makes visually apparent the connection between the sine of an angle represented by the unit circle and its representation on a Cartesian coordinate plane. Moreover, the sketch also allows one to demonstrate trigonometric identities such as sin ($\pi/2 + \theta$) = cos θ (where θ is

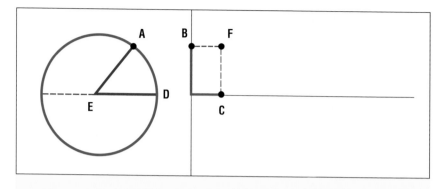

Fig. 10.12. Coordinating the length of arc *DA* and verticle position of point *A*

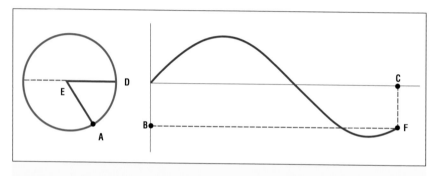

Fig. 10.13. Showing the correspondence as a continuous function

the measure of angle *DEA*) by placing point *A* at an angle of $\pi/2$ and then moving point *A* along the circumference.

We now return briefly to the question from the opening of this section: Why does a graph of the sine function behave as it does? In dynamically linking the two models—the unit circle and the Cartesian coordinate plane—teachers are better positioned to engage their students in a conceptual conversation about the factors that underlie each model and thus help explain the behavior of the graph of the sine function.

Concluding Remarks

The three examples presented in this article illustrate how technology can be used to support teachers' efforts to foster students' understandings and intuitions by engaging them in conceptual conversations. In the systems of equations example, technology offers teachers the ability to unpack the many mathematical relationships that underlie a graphical representation of a system of equations. These relationships include the correspondence between two variables in a functional relationship and the quantitative comparison of the two functions in the system. The ability of interactive geometry programs to hide and show objects and their dynamic dragging features encourage classroom conversations about each relationship in turn, and thus they promote a deeper understanding of the graphical solution itself.

The line-of-best-fit example further illustrates the capabilities of interactive geometry software to provide enhanced images that promote conceptual conversations. In this example, images of the "least square," constructed in an interactive geometry environment, allow for real-time reflection on the fit of a line to a set of data. This sketch can be used to enhance the popular instructional approach of using a spaghetti "straightedge" to estimate a line of best fit. By providing this capability, interactive geometry programs allow teachers to illustrate the mathematical objects behind the concept of a least squares regression by using dynamic images to foster engagement with a mathematical concept that is often explained superficially, if at all.

The final example, the sine function behavior example, illustrates a capability that is unique to technologies, computer or otherwise—dynamically linked objects. By dynamically linking two different mathematical objects, a point on a circle and a point on a Cartesian coordinate plane, this example portrays the sine function in a way that allows students to observe and describe the related motions of the points during a conceptual conversation. In this example, we highlight the use of technology to generate explanations of a mathematical relationship through a classroom conversation.

Although this article has focused specifically on the use of a particular technology, the ideas and illustrations are meant to be representative of the ways in which technology more generally can provide a powerful means of fostering students' understanding and intuition of the mathematics they study. In describing each example, we deliberately highlight one aspect of a conceptual conversation—relationships, images, or explanations—in order to illustrate the variety of ways in which technology might be used to engage students in such conversations. We believe, however, that a good conceptual conversation addresses each of these aspects, and we conclude by encouraging the reader to explore the three Geometer's Sketchpad demonstrations in the CD accompanying this yearbook and to consider the role that relationships, images, and explanations play in acquiring a deeper understanding of the mathematical concepts in each example.

REFERENCES

Bennett, Dan. *Exploring Geometry with the Geometer's Sketchpad.* Emeryville, Calif.: Key Curriculum Press, 1999.

Thompson, Patrick. "Imagery and the Development of Mathematical Reasoning." In *Theories of Mathematical Learning,* edited by Leslie Steffe, Paula Nesher, Paul Cobb, Gerald Goldin, and Brian Greer, pp. 267–83. Mahwah, N.J.: Lawrence Erlbaum Associates, 1996.

11

The Spreadsheet: A Vehicle for Connecting Proportional Reasoning to the Real World in a Middle School Classroom

Richard Caulfield
Paul E. Smith
Kelly McCormick

A s INDICATED in the *Principles and Standards for School Mathematics* (National Council of Teachers of Mathematics [NCTM] 2000), technology needs to play an important role in the teaching and learning of mathematics. Technology can be used as a vehicle to extend the range and quality of investigations by allowing students to focus on decision making, reflection, reasoning, and problem solving. The purpose of this article is to present two rich examples of how the spreadsheet can be used in the middle school mathematics classroom as a means of connecting mathematics to contexts outside of mathematics. The first activity will show the usefulness of the spreadsheet when working with and manipulating real-world numbers. The emphasis of this activity will be on proportional reasoning and making mathematical connections with social studies. The second activity uses the power of the spreadsheet to improve middle school students' understanding of the probability of a real-world event: the odds of winning the lottery.

The States Spreadsheet Investigation

In launching this activity, the teacher handed students a blank map of the United States and asked them to work in groups of three or four to identify the ten largest states in size by area by placing the numbers 1 (largest) through 10 (tenth largest) in the appropriate state on the map. The groups all shared what they believed to be the top four states in area. They were then given a spreadsheet file containing the area, population, and number of representatives in the U.S. House of Representatives for each state (see table 11.1 and the accompa-

---

Editor's note: The CD accompanying this yearbook contains Word documents and Excel files that are relevant to this article.

nying file on the CD). The students used the spreadsheet to sort the states by their actual area from largest to smallest. Even though Alaska was the largest state, none of the groups had originally chosen Alaska as the largest in area. In fact, one group did not even include Alaska in its list. The students were asked to explain why Alaska had been left out as their first choice for the largest state. After some debate, a group volunteered that Alaska was drawn to a different scale. When asked why this was done, the students explained that if the states were all drawn to the same scale, Alaska wouldn't fit on the page.

Table 11.1
United States Information Spreadsheet

State	Area (sq.mi.)	Population	Representatives
Alabama	52419	4447100	7
Alaska	663267	626932	1
Arizona	113998	5130632	8
Arkansas	53179	2673400	4
California	163696	33871648	53
Colorado	104094	4301261	7
Connecticut	5543	3405565	5
Delaware	2489	783600	1
District of Columbia	68	572069	0
Florida	65755	15962378	25
Georgia	59425	8186453	13
Hawaii	10931	1211537	2
Idaho	83570	1293953	2
Illinois	57914	12419293	19
Indiana	36418	6080485	9
Iowa	56272	2926324	5
Kansas	82277	2688419	4
Kentucky	40409	4041769	6
Louisiana	51840	4468976	7
Maine	35385	1294464	2
Maryland	12407	5296486	8
Massachusetts	10556	6349097	10
Michigan	96716	9938444	15
Minnesota	86939	4919479	8
Mississippi	48430	2844658	4
Missouri	69704	5595211	9
Montana	147042	902195	1
Nebraska	77354	1711263	3
Nevada	110561	1998257	3
New Hampshire	9350	1235766	2
New Jersey	8721	8414350	13
New Mexico	121589	1819046	3
New York	54556	18976457	29
North Carolina	53819	8049313	13
North Dakota	70700	642200	1

Table 11.1 (*continued*)
United States Information Spreadsheet

State	Area (sq.mi.)	Population	Representatives
Ohio	44825	11353140	18
Oklahoma	69898	3460097	5
Oregon	98381	3421399	5
Pennsylvania	46055	12287150	19
Rhode Island	1545	1048319	2
South Carolina	32020	4012012	6
South Dakota	77116	754844	1
Tennessee	42143	5689283	9
Texas	268581	20851620	32
Utah	84899	2233169	3
Vermont	9614	608827	1
Virginia	42774	7078515	11
Washington	71300	5894121	9
West Virginia	24230	1808344	3
Wisconsin	65498	5363675	8
Wyoming	97814	493782	1
Total	3794086	281436777	435

World Almanac Books. *The World Almanac and Book of Facts 2003.* New York: World Almanac Education Group, 2003

The class was then told to estimate how much larger the paper would have to be to draw Alaska to the same scale. They first responded with just guesses, but when asked to support their guesses with mathematics, they began to explore the scale of the map and the insert of Alaska. Using a map from the class atlas, one student discovered that Alaska was drawn to a scale of 2.2 cm = 400 miles, whereas the rest of the United States (excluding Hawaii) was drawn to a scale of 2.5 cm = 150 miles. The different groups then had discussions about how to compare these figures, and most of the groups eventually decided that they should find out how many miles were represented by 1 centimeter for each drawing. After working for several minutes, a group went to the board and explained that they figured out that 1 centimeter represented about 182 miles for the Alaskan insert, whereas 1 centimeter represented about 60 miles for the U.S. map. The group concluded that it would take a sheet three times bigger to include Alaska at the same scale. When pressed if this was the smallest piece of paper that would work, the group quickly said yes, but then they quickly withdrew their yes and said that they weren't sure.

Next, the students were asked to create a graph comparing the largest four states (see fig. 11.1). Using the spreadsheet, some created histograms; others made box plots, but one group created a circle graph. The groups were then

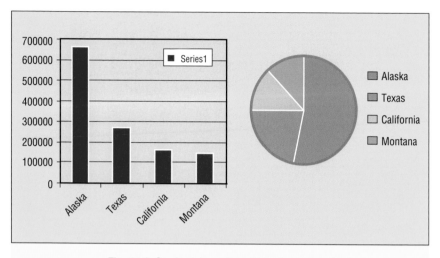

Fig. 11.1. Students' graphs of four largest states

asked to evaluate the value of the different representations. To help them with this task, they were told to look at the types of conclusions that they could draw from each of the graphs. After some discussion, the students concluded that although all the graphs represented the overall size of Alaska compared to the other three states, the circle graph's representation made obvious the fact that Alaska is larger in area than the other three states combined.

For their next task, the students were asked to find the percent of area for the entire United States that would be found within the borders of Alaska. They decided that they could find the area of the United States by taking the sum of the column in table 11.1 showing the area of each state and then dividing the area of Alaska by the total area of the United States. That would give the percent. From this, the students discovered that Alaska occupied about 17.5 percent of the total area of the United States.

During the next phase of the lesson, the students were told that the majority of land in Alaska was actually owned by the United States government. Some people in the United States and many in Alaska would like to drill for oil on U.S. government land that happens to be in Alaska. If a majority of Alaskans wanted to drill for oil, which might provide jobs for Alaskans, what chance would they have of convincing the government of the United States that drilling should take place? To answer this question, the students determined the proportion of representatives in the House who were from Alaska. After discovering that only 1 out of the 435 representatives were from Alaska, the students recognized the difficulty Alaska would have in arguing its case before the House. This led to a discussion centered on the creation of the two houses of Congress from a historic perspective, recognizing the founders' interest in helping less populated states have a more equal say in federal issues with the creation of the Senate.

The same process was followed as we looked at population. Students sorted the states by population, and graphs were made of the largest four states in population. The groups determined that a histogram and a box plot were just as effective as a circle graph in representing these particular data. They conjectured that the data being displayed determined the appropriateness of the type of graph to be used. However, when pressed further to see if anything else would be a factor in the appropriateness of the type of graph they should use to display data, several students decided that the audience to whom the graph was being presented would also play a role, as would the type of information being presented.

The lesson concluded with the students being assigned homework. They were told to use the data on the spreadsheets to develop a mathematical conjecture on how the number of representatives to the House of Representatives for each state is determined. They defended their conjectures during the next day in class. One of the students stated that he took the total population of the state and divided by the total number of representatives in the House. He then had the spreadsheet calculate the average of those figures and suggested that the states received a representative for about every 670,000 people. However, someone pointed out that South Dakota had only one representative with a population of 756,600. Someone else explained that she ordered the states from smallest to largest and attempted to identify the populations that allowed for an additional representative. Another student explained that she used a proportion comparing the size of the population of a state to the number of representatives. She had decided on the following proportion: a state's population / total (U.S.) population = the number of a state's representatives / total number of representatives (435). She had also developed a formula that would apply this theory. She divided the state populations by the total population and then multiplied by 435. When she formatted the spreadsheet columns to round to the nearest whole number, the results were the closest yet to the actual number of representatives in the House.

A good mathematical task demands communication and reflection (Hiebert et al. 1997), and the key to this lesson was that it gave way to rich discussions and reflections. Thus, the spreadsheet was a valuable tool; the students focused on conjecturing, experimenting, communicating, and reflecting on real-world data without cumbersome computations getting in their way. Throughout this activity, the students were able to develop a sense of proportion and how proportional reasoning is an important mathematical concept outside the school environment.

What Are the Odds? Simulating the Lottery

Many states have a lottery, and many people play the lottery without knowing the tremendous odds against winning. Others know the odds but have no tangible way to understand the magnitude of the odds. This activity brings to light the real odds and the reality of playing the lottery.

Simulation is a great way to have students discover the "realness" of a probability event. When those probabilities are small, coins and dice can be used in simulation. However, the chances of winning most state lotteries are in the tens of millions, and so it is going to be impossible to use coins or dice or to pull numbers out of a bag to simulate playing the lottery. Spreadsheets offer a solution to this dilemma. In addition to crunching numbers quickly, spreadsheets can randomly generate numbers and compare them to another set of numbers that the user selects in a matter of seconds.

In this activity, the class began with a discussion of the state lottery system. The students were asked the following questions: "Does it pay to play the lottery? Do you stand a good chance of winning? What are the odds of winning? How many tickets would you have to buy in order to have a good chance of winning? How could you express the probability of winning the lottery, and what do you think is the magnitude of that proportion?"

After discussing these questions, the students were told that they were actually going to play a simulation of the lottery on their computers. The "Lottery Simulation" (see table 11.2 and the accompanying file on the CD) simulates a Pick Six (numbers) game, sometimes called a "6:10" game, and allows the teacher to control the expanse of the simulation. For the first simulation, the students were to pick six numbers from 1 to 10. They actually thought that their chances of winning would be pretty good if this were the situation for real lotteries. "Lottery Simulation" allows the student to play fifteen separate lottery cards each time they hit the F9 key (they should be reminded that this would be $15 worth of playing). They were permitted to pick another set of six numbers at any time during the exercise. They were not limited in the number of their attempts, though they were required to keep track of the number of times they "played the Lotto." To see if they had won, they checked the column at the far right. If a "6" appeared followed by the word "OK," that meant they had matched all six numbers and had won the lottery! A "3" represented three matches, and so on.

Toward the beginning of the simulation, we discussed how the simulation differed from the real lottery. One student brought up the fact that in real life when people actually buy a ticket, they have the option of letting the computer pick their numbers. We also talked about how the numbers picked in the real lottery must all be different. In "Lottery Simulation," occasionally they

Table 11.2
Lottery Simulation Spreadsheet

	A	B	C	D	E	F	G	H	I	J	K	L	M	N	O	P
1				**Lottery Simulation**												
2																
3				Possible numbers to choose from can be placed here →			50									
4			Lottery Number Computer Picks					Players can choose any six numbers in bold, then press F9 to play							Winner = 6 and OK	
5	1st	2nd	3rd	4th	5th	6th		22	21	17	23	8	19		↓	
6	33	25	30	30	20	39		0	0	0	0	0	0		*0*	invalid
7	13	17	22	4	8	45		1	0	1	0	1	0		*3*	OK
8	37	44	16	41	3	25		0	0	0	0	0	0		*0*	OK
9	2	22	4	6	12	17		1	0	1	0	0	0		*2*	OK
10	14	40	43	29	28	43		0	0	0	0	0	0		*0*	invalid
11	33	22	13	42	19	11		1	0	0	0	0	1		*2*	OK
12	13	7	37	4	41	35		0	0	0	0	0	0		*0*	OK
13	21	44	31	22	9	35		1	1	0	0	0	0		*2*	OK
14	12	8	33	17	21	17		0	1	1	0	1	0		*3*	invalid
15	15	12	8	17	39	31		0	0	1	0	1	0		*2*	OK
16	14	29	2	15	1	11		0	0	0	0	0	0		*0*	OK
17	5	25	1	10	31	9		0	0	0	0	0	0		*0*	OK
18	37	38	30	13	13	1		0	0	0	0	0	0		*0*	invalid
19	34	39	10	1	1	36		0	0	0	0	0	0		*0*	invalid
20	12	30	41	3	10	3		0	0	0	0	0	0		*0*	invalid
21	32	28	26	38	10	38		0	0	0	0	0	0		*0*	invalid

↑
This section randomly generates numbers similar to a computer pick in the lotto game.

↑
This section checks your picks against the computer picks.

were not. So in order to ensure the validity of the simulation, the number had to be followed by an "OK." If the numbers chosen were in fact repetitive, an "invalid" would instead appear.

Occasionally, the students would hit the "jackpot" in the simulation of a "6:10" Lotto game. This led to a great discussion about the proportional reasoning behind this probability event. We talked about the probability of winning a "6:10" Lotto game, 1/210, and what this probability meant regarding their playing and winning.

Next, the students were told that they were going to play a simulation of the lottery that reflected the actual Pick Six game in either their state or on their computers, and so instead of picking six numbers from 1 to 10, they were to pick 6 numbers from 1 to 50. The students were eager to see if someone could again win; however, they soon realized that they were wasting their time (and money) picking the numbers.

The students were given ten minutes to try and win. After the ten-minute period, the class was brought back together to discuss what just happened. The class calculated how much everyone had just "spent" playing the lottery, which ended up being about $45,000 in ten minutes! The students were then asked to express as a proportion the amount of money that was spent every minute in this simulation. We also discussed the actual probability of winning a "6:50" game, 1/15,890,700, and what this probability meant in regard to their playing and winning. For homework the students were to write about the following two questions: How would the probability of winning change if the initial simulation was changed to a "5:10" Lotto game? How would the probability of winning change if the initial simulation was changed to a "7:10" Lotto game?

The students came away from this lesson with a greater understanding of proportional reasoning in dealing with large probabilities. The NCTM (2000) *Principles and Standards* stresses that middle school students "need to develop their probabilistic thinking by frequent experience with actual experiments" and that "computer simulations may help … avoid or overcome erroneous probabilistic thinking" (p. 254). "The importance of probability questions in the context of real data supersedes past approaches … that started either with counting problems or with games of chance" (Shaughnessy 2003, p. 224). Thus, the spreadsheet again proved to be a valuable tool; through the lottery simulation, the students gained a new appreciation for the probability of a real-world event, winning the lottery.

Technology Enhances Learning

The *Principles and Standards* (NCTM 2000) recommends that students in grades 6–8 should develop facility with proportionality through working in many areas of the curriculum, including ratio and proportion, percent, scaling, and probability. "The understanding of proportionality should also emerge through problem solving and reasoning, and it is important in connecting mathematical topics and in connecting mathematics and other domains" (p. 212). "In grades 6–8, all students should use proportionality and a basic understanding of probability to make and test conjectures about the results of experiments and simulations" (p. 248). In both of these activities, the students were working to develop their proportional reasoning, and the technology served as a vehicle through which the students were able to formulate, explore, and evaluate conjectures. In the first activity, the technology provided a focus around which the students discussed with one another and their teacher the objects on the screen. During the second activity, the lottery simulation allowed students to experience a problem situation that would have been impossible to create without the use of technology. The simulation proved to be a valuable classroom tool for making a somewhat abstract probability concept more concrete. In both these activities, the spreadsheet proved to be a rich medium that facilitated students' learning about mathematical contexts in the real world.

REFERENCES

Hiebert, James C., Thomas P. Carpenter, Elizabeth Fennema, Karen C. Fuson, Diana Wearne, Hanlie Murray, Alwyn Oliver, and Piet Human. *Making Sense: Teaching and Learning Mathematics with Understanding.* Portsmouth, N.H.: Heinemann, 1997.

National Council of Teachers of Mathematics (NCTM). *Principles and Standards for School Mathematics.* Reston, Va.: NCTM, 2000.

Shaughnessy, J. Michael. "Research on Students' Understandings of Probability." In *A Research Companion to "Principles and Standards for School Mathematics,"* edited by Jeremy Kilpatrick, W. Gary Martin, and Deborah Schifter, pp. 216–26. Reston, Va.: National Council of Teachers of Mathematics, 2003.

12

Using Spreadsheet Software to Explore Measures of Central Tendency and Variability

Carol S. Parke

THE study of data analysis, statistics, and probability is essential for many reasons. We come into contact with statistics nearly every day of our lives. We hear results from political polls that tell us the state of the nation or results from marketing studies that tell us the most popular brand of a product. Radio, television, newspapers, and magazines report data on economic, social, and political issues relevant to our country; and we learn about medical research that identifies risk factors associated with certain diseases. As educators, we need to ensure that our students become knowledgeable, informed, and critical consumers of this research. Moreover, several career opportunities for students in natural science, mathematics, engineering, political science, social science, testing, education, and economics often incorporate statistical concepts.

One of the first steps toward empowering students with statistical knowledge is to provide ample opportunities for them to develop data sense. Decisions must be made about the best ways to summarize raw data and the most appropriate ways to interpret the data. As students develop understandings of data representation and summarization, they are building data sense and are considering the "when and why" of using statistics (Friel 1998). Students must "think about (1) what each measure tells them about the data and (2) how this information will help in the final stage of interpreting and communicating the results" (p. 215).

In *Principles and Standards for School Mathematics* (National Council of Teachers of Mathematics [NCTM] 2000), the Data Analysis and Probability Standard states that students should be able to describe data to answer questions, select appropriate statistical methods, make predictions, and apply probability concepts. The illustrations provided here focus on the selection

---

Editor's note: The CD accompanying this yearbook contains an Excel file that is relevant to this article.

and use of appropriate statistical concepts to describe data. One category of descriptive statistics is often referred to as *measures of central tendency.* They convey information about the center of the distribution and yield a value that best represents the typical score. The most common of these statistics are the mean, median, and mode. These measures tell only one part of the data story, however. Another category, called *measures of variability,* is essential both in determining how "spread out" or diverse the values are within the data set and in describing the degree to which the data are homogeneous or heterogeneous. Common measures include the range, interquartile range, and standard deviation. When equipped with a firm grasp of the underlying concepts of these descriptive statistics, students can attain a more complete picture of the data.

Much research has been conducted on how students learn about mean, median, and mode as well as the understandings or misunderstandings they have about these measures. Mokros and Russell (1995) studied how children think about the mean as a mathematical definition. Students from fourth to eighth grade were asked to think aloud as they solved open-ended problems. Some students considered the mean as simply an algorithmic procedure, whereas others viewed it more conceptually as a mathematical point of balance. Strauss and Bichler (1988) investigated the development of children's concepts of average. Seven properties of the mean were identified, including those related to the mathematical function itself (e.g., the mean is influenced by every value) and the abstract nature of the mean (e.g., the mean may not necessarily equal any of the values). Finally, in a study examining solution strategies to an average problem, Cai (1998) states that a conceptual understanding of average should include an understanding of the computational algorithm as well as the statistical nature of the concept.

Enhancing Statistical Learning Using Spreadsheet Software

This paper demonstrates how a spreadsheet program enables students to gain a conceptual understanding of statistical measures and use them to describe data sets more fully. In particular, measures of central tendency (mean, median, and mode) and variability (range and standard deviation) are explored. Incorporating technology into data analysis instruction alleviates the necessity to perform statistical calculations repeatedly. Values in a data set can easily be added, deleted, or modified in a program and the statistics recalculated, thereby allowing students to speculate on the changes that occur and the reasons *why* they occur.

Using technology in data analysis instruction also enhances the level of mathematics that can be taught. When requesting descriptive statistics in spreadsheet programs, an abundance of measures is usually produced, thus exposing students to these concepts at an early age. For instance, the concept of standard deviation can be examined and compared across data sets even though students may not have had formal instruction on the algorithm or procedures for its computation. Typically, less attention in mathematics instruction is given to the standard deviation because the formula may seem somewhat formidable to children because of its computational complexity.

Computer software, such as Microsoft Excel, provides a wealth of statistical information. In addition, many statistical packages, such as SPSS for Windows, Stat View, S-Plus, and Minitab, are relatively inexpensive and user-friendly. Some handheld devices will also produce the results that follow (e.g., the TI-83, TI-89, TI-92 Plus, and TI Voyage 200).

Exploring Central Tendency and Variability Concepts

The sections below describe a lab activity in which students in Mrs. Remille's mathematics classroom used a spreadsheet program to compare and contrast data sets and their statistical properties. Prior to the activity, her students had received instruction on the mean, median, mode, and range. They learned the algorithmic formulas, they practiced finding these measures on several data sets, they created bar graphs and other visual displays of data, and they were given opportunities to solve open-ended problems that demonstrated the use of statistics in real-life situations. With regard to prior exposure to technology, her students had some experience with computer software in other courses within their school, and in Mrs. Remille's mathematics classroom, students were comfortable using a calculator. This lab activity, however, was these students' first exposure to using a spreadsheet program to learn about statistics.

As Mrs. Remille was creating the activity, she thought about how she wanted her students to become involved with the data. She decided to begin with one data set consisting of twenty assessment scores. This "original" data set would serve as a reference point for subsequent comparisons. She knew the statistical properties she wanted to bring to the forefront, and also one of her goals was to introduce the concept of standard deviation. Therefore, she wanted to ensure some amount of systematic progression throughout the session. On the one hand, she knew there would be times when she would tell the students to "do this to the data and find out what happens." On the other hand, Mrs. Remille also wanted her students to have some freedom to try their own changes to the data. She wanted the decision of "what to do next" to be guided in part by the discussion that would ensue during the session. She thought this idea was important to motivate her students and make the learning interesting for them. It is interesting to notice that as the activity unfolded, some steps were guided by the teacher and others were initiated by students.

Examining the "Original" Data Set

Mrs. Remille first showed students how to enter the twenty data points into a column in the spreadsheet and then showed them how to "sort" the values from lowest to highest and request descriptive statistics. The data are shown in the first column in table 12.1, labeled as "original." (Note that the data are also available in electronic format on the CD accompanying this yearbook.) It took a bit of time for students to become familiar with the software initially, but with each subsequent change to the data, they became more proficient in its use.

Table 12.1
Spreadsheet Containing "Original" Data Set and Subsequent Modifications

original	distance	data1	distance1	data2	data3	data4	data5	data6	data7
15	-4	11	-8	0	15	15	20	20	18
15	-4	12	-7	15	15	15	20	20	18
16	-3	14	-5	16	16	16	21	20	18
16	-3	14	-5	16	16	16	21	20	18
17	-2	15	-4	17	17	17	22	20	18
17	-2	15	-4	17	17	17	22	20	18
18	-1	16	-3	18	18	18	23	20	18
18	-1	17	-2	18	18	18	23	20	18
19	0	17	-2	19	19	19	24	20	18
19	0	19	0	19	19	19	24	20	18
19	0	19	0	19	19	19	24	20	22
19	0	19	0	19	19	19	24	20	22
20	1	20	1	20	20	20	25	20	22
20	1	22	3	20	20	20	25	20	22
20	1	22	3	20	20	20	25	20	22
21	2	23	4	21	21	21	26	20	22
21	2	24	5	21	21	21	26	20	22
22	3	25	6	22	22	22	27	20	22
23	4	27	8	23	23	23	28	20	22
25	6	29	10	25	25	25	30	20	22
					19	25			

Table 12.2 shows the statistical information for the "original" data that Mrs. Remille asked students to produce. Most software packages will allow the user to select the statistical measures to be included in the output. Other statistics that may be available, but not included in these illustrations, are standard error, skewness, and kurtosis.

Table 12.2
Descriptive Statistics for the "Original" Data

original	
Mean	19
Median	19
Mode	19
Standard Deviation	2.64
Sample Variance	6.95
Range	10
Minimum	15
Maximum	25
Sum	380
Count	20

One student immediately noticed that the mean, median, and mode were equal and wanted to look at the data visually, so the class created a bar chart (fig. 12.1). In looking at the distribution, he said, "I can tell that 19 is the most common score on the test; it's the average of all the scores, and there's an equal number of people above and below this score." Another student noticed that the range was 10 and that "no one got above 25 or below 15." Some students also questioned the unfamiliar numbers in the output. Mrs. Remille took this opportunity to discuss how 19 is a good representation of the data according to all three measures of central tendency but that scores on the assessment varied. She introduced the notion of standard deviation by asking students to find out how far away each score was from the mean. The second column in table 12.1, labeled "distance," shows the distance and direction of each score from the mean of 19. Students looked at this column of numbers and noticed that "four numbers are 0 because the scores were the same as the mean" and that "6 is the biggest distance above and 4 is the biggest distance below the mean." The class talked about what the average distance would be, and the teacher referred them to the number 2.64 in the output. Thus, as students examined the values in the "distance" column, they were developing a beginning understanding of the standard deviation as an average distance of scores from the mean.

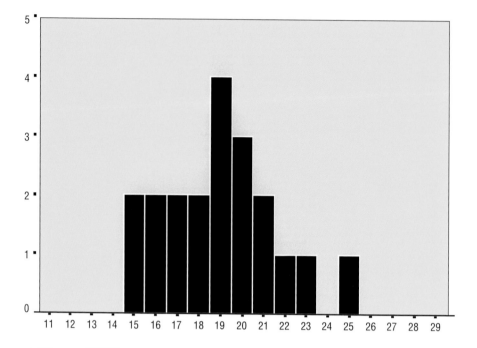

Fig. 12.1. Bar graph of the "original" data set

Comparing Data Having a Higher Degree of Variability

Mrs. Remille then asked students to enter a new set of twenty assessment scores (see the third column of table 12.1, labeled "data1") and produce the descriptive statistics (table 12.3). Students immediately recognized that the mean, median, and mode were the same as the "original" data, but a few of the other measures were very different. One student said, "The range is a lot bigger; here it's 18, but in the other set it was only 10." The teacher directed students' attention toward the standard deviation (4.98) and asked why it was larger. Students were not quite sure how to answer the question, so they created a bar graph for the new data, as shown in figure 12.2.

The teacher and students talked about how the data in figure 12.2 looked more "spread out" than the data in figure 12.1. As described by one student, "The scores in the first group were more bunched up closer to 19 [the mean], but in the next group a lot of them were farther away." This comparison served as a good illustration of how two data sets can have similar measures of central tendency but be quite different in the variability or deviation of the scores. They also discussed how the mean is less representative of the data in

Table 12.3
Descriptive Statistics for "data1"

	data1
Mean	19
Median	19
Mode	19
Standard Deviation	4.98
Sample Variance	24.84
Range	18
Minimum	11
Maximum	29
Sum	380
Count	20

figure 12.2 compared to figure 12.1. Finally, Mrs. Remille asked her students to create a column that listed the distance of every score to the mean (see the fourth column, "distance1," in table 12.1). Students commented on how much larger these numbers were in comparison to the distances for the original data,

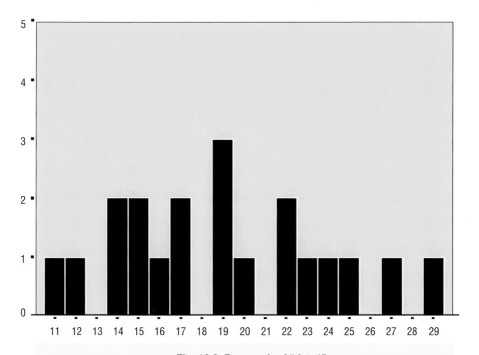

Fig. 12.2. Bar graph of "data1"

and the class talked once again about how the standard deviation represented the average distance of scores to the mean.

Modifying One Value in the "Original" Data to Become an Outlier

The teacher asked her class to refer to the original data and find out what happens to the statistics if one of the scores that was equal to 15 is changed to 0. The modified data set is shown in the "data2" column in table 12.1, and the descriptive statistics are displayed in table 12.4. The effects of changing a score to become an outlier, or extreme value, are most evident in the reduction of the mean and the increase in magnitude of the standard deviation. Before getting the results, one student predicted that the mean, median, and mode would change. He was surprised to discover that only the mean changed. Another student explained that "19 is still the most common score, the mode; and 19 is still the median because 15 is just replaced by 0 so there are still half above and half below." The teacher asked why the mean changed, and the student responded, "It got lower because one of the kids who got 15 in the first example got a 0 the next time around, so when you add up all the new scores and divide by 20, the answer [the mean] is less." The class then looked at the "Sum" value given in the output that provides the total of all scores and noticed that it was 15 points less than the "Sum" value in the original data. The class also discussed how the mean can be a decimal value (18.25) even though none of the values in the data are decimals. To examine more data sets, students changed another existing score to either a very low or a very high value. Although not shown here, the results varied, and students were able to realize that depending on what score they changed and what value they changed it to, the mean, median, and mode may or may not be affected.

Table 12.4
Descriptive Statistics for "data2"

	data2
Mean	18.25
Median	19
Mode	19
Standard Deviation	4.95
Sample Variance	24.51
Range	25
Minimum	0
Maximum	25
Sum	365
Count	20

Finally, they compared the new standard deviation, 4.95, and noticed that it was much higher than the standard deviation for the original data. The class discussed how one extreme value can have a great impact on this measure of variability. The new score of 0 was 18.25 points below the mean, whereas in the original data, the score of 15 was only 4 points below. Although not shown here, bar graphs for these data were created and compared to the previous ones. A student noticed that the standard deviations for "data1" and "data2" were almost the same, but the graphs looked very different. This sparked further discussions about the effect of outliers.

Adding Values to the "Original" Data Set

Another alteration of the "original" data answered the question, What happens when a score is added to the data? Mrs. Remille asked students to suppose that one child was absent during the original assessment and when he took it later, his score was equal to the original mean of 19. This new data set is represented in the "data3" column in table 12.1, and the statistical output is shown in table 12.5. Students were asked to predict the effect of this additional value before they actually computed it. The class discussed how the mathematical point of balance, the mean, did not change when a score equal to the mean was added. The mode and median also remained at 19, but there was a slight effect on the standard deviation. One student asked, "Why would it be lower? The distance between the new score of 19 and the mean of 19 is 0, so it shouldn't change." This question was an opportunity to further students' understanding of the concept of standard deviation. Although the calculated distance between scores and the mean did not change (i.e., the numerator of standard deviation), there are now twenty-one scores instead of only twenty

Table 12.5
Descriptive Statistics for "data3"

	data3
Mean	19
Median	19
Mode	19
Standard Deviation	2.57
Sample Variance	6.60
Range	10
Minimum	15
Maximum	25
Sum	399
Count	21

scores, which affects the denominator of standard deviation. Thus, the average distance between the scores and the mean is slightly reduced.

When the new score is farther from the mean, different changes occur. Next, students changed the absent child's score to 25 instead of 19 (as shown in the "data4" column in table 12.1). Results in table 12.6 indicate that the mean and the standard deviation are slightly larger than the original values. Students discussed why these statistics were larger. They also predicted what would happen if the new score was 6 points below the mean (13) instead of 6 above (25).

Table 12.6
Descriptive Statistics for "data4"

	data4
Mean	19.29
Median	19
Mode	19
Standard Deviation	2.88
Sample Variance	8.31
Range	10
Minimum	15
Maximum	25
Sum	405
Count	21

Increasing Every Value in the "Original" Data Set by a Constant

To explore another characteristic of the mean and standard deviation, the question posed was, What happens if 5 points are added to every child's original assessment score? The modified data are in the "data5" column of table 12.1, and the resulting statistics are shown in table 12.7. Each of the measures of central tendency is increased from 19 to 24, a difference of 5 points. At this point, the teacher asked students to create additional examples that added (or subtracted) differing amounts to each original data value. Students shared with the class what happened to their statistics. They recognized that the mean increased (or decreased) by the same amount that was added (or subtracted) to each score. An observant student also noticed that the value in the "Sum" column, which is the numerator for the mean, increased by 100 "because I added 5 to every score and there were twenty scores, so I really added 100 total points but the number of kids was still 20, so the mean was 480 divided by 20, which is 24."

Table 12.7
Descriptive Statistics for "data5"

	data5
Mean	24
Median	24
Mode	24
Standard Deviation	2.64
Sample Variance	6.95
Range	10
Minimum	20
Maximum	30
Sum	480
Count	20

Another student commented that the new standard deviation was exactly the same as the original standard deviation of 2.64. Students created a new column that showed the distances between each score and the mean. Although not shown here, it demonstrated that when each individual value is increased or decreased by the same amount, the average distance between the scores and the mean remains unchanged. Visually, bar graphs of the new data and original data look the same. This alteration of data had the effect of simply shifting the entire distribution up (or down) the scale a given number of points; the distances between the values remained the same.

Analyzing a Set of Constant Values

This analysis came about when one of the students was curious about what would happen if all the children received the same score of 20 on the assessment (as shown in the "data6" column in table 12.1). After examining the results in table 12.8, one student pointed out that "almost every one [measure] is 20," and another student said, "There is no standard deviation." Perhaps the most interesting statistic to examine here is the standard deviation, which indicates that there is no variation in the data. Students discussed why the value is 0. As a student explained, "The mean is 20 and every score is 20, so the distance between each score and the mean is 0. It wouldn't matter how many scores there were. If they were all 20, then the standard deviation would still be 0."

Instead of all scores being a constant value, another student wondered, "What would happen if half of the scores are 18 and the other half of the scores are 22?" These data are given in the "data7" column of table 12.1. Several interesting statistics were discussed (see table 12.9). A student questioned, "Why would the mean and median be 20, but no one scored a 20?"

Table 12.8
Descriptive Statistics for "data6"

	data6
Mean	20
Median	20
Mode	20
Standard Deviation	0
Sample Variance	0
Range	0
Minimum	20
Maximum	20
Sum	400
Count	20

Another student responded that "it doesn't matter that nobody had a 20. When you add up 18 ten times and 22 ten times, you get the 'Sum' of 400, and 400 divided by 20 is 20. The median is 20 because it's the number between 18 and 22." This illustrates a statistical property of mean and median; that is, these two measures can have values that are not obtained by anyone in the sample data set. The class also discussed the mode of 18. Certain software packages may produce different results for the mode in some situations. Here, the data are actually bimodal (modes of 18 and 22); however, other software may indicate that there is no mode, may display all modes, or may indicate the smallest or largest modal value.

Table 12.9
Descriptive Statistics for "data7"

	data7
Mean	20
Median	20
Mode	18
Standard Deviation	2.05
Sample Variance	4.21
Range	4
Minimum	18
Maximum	22
Sum	400
Count	20

Depending on students' level of mathematical sophistication, the value for the standard deviation may also be beneficial to discuss. Ten scores are 2 points above the mean and ten scores are 2 points below the mean. Conceptually, the standard deviation is the average distance between the scores and the mean. All scores are a distance of 2 points from the mean, but the standard deviation is reported to be 2.05 instead of 2.00. Most spreadsheet software programs calculate the sample standard deviation, which divides the distances by $n - 1$, rather than the population standard deviation, which divides the distances by n, thus leading to a slightly larger value. At any rate, the two values are usually very similar.

Conclusion

There are numerous and varied benefits of using technology in statistics instruction. When students have opportunities to examine and compare analyses of several data samples, they develop a deeper understanding of concepts that form the basic foundation of statistical methods, such as central tendency and variability. Without becoming bogged down by the calculations involved, students begin to discover patterns of change in the statistical measures when data are altered. This can lead to a better appreciation of the underlying meaning of the formulas. Students also learned a new concept of standard deviation that can be quite daunting at first when focusing only on the algorithmic procedure.

Students in Mrs. Remille's classroom became increasingly engaged as the lab activity progressed. In part, this was due to the way she allowed the session to unfold. Although she had in mind a certain series of data changes that she wanted her students to make, she was not rigid in how they played out during the activity. She often gave students the freedom to decide the avenues they would explore. There were several instances in which students asked, "What would happen if...?" and thus she incorporated their suggested data changes into the activity. In essence, she was creating an environment that got students motivated, encouraged them to think at a deeper level, and involved them in important statistical discussions.

The description of the lab activity presented in this paper illustrates only a small portion of the statistical learning that can take place when students become familiar with spreadsheet software and its capabilities. As you try this activity with your own students, you may choose not to go through the specific changes shown here but instead let the data alterations progress naturally as a result of your own unique class discussion. As students progress in their study of mathematics, more-advanced statistical procedures can be explored—for example, inferential statistics that allow for making generalizations from sample data to a population. The spreadsheet program used in Mrs.

Remille's classroom could also provide statistics such as skewness and kurtosis to describe the shape of the data (e.g., skewed, normal, uniform). These illustrations are meant to furnish a starting point for students' exploration of data. Students can come up with their own modifications to the data, make predictions, and perform analyses to determine the outcomes. The possibilities are virtually endless.

REFERENCES

Cai, Jinfa. "Exploring Students' Conceptual Understanding of the Averaging Algorithm." *School Science and Mathematics* 98 (February 1998): 93–98.

Friel, Susan N. "Teaching Statistics: What's Average?" In *The Teaching and Learning of Algorithms in School Mathematics: 1998 Yearbook,* edited by Lorna J. Morrow, pp. 208–17. Reston, Va.: National Council of Teachers of Mathematics, 1998.

Mokros, Jan, and Susan Jo Russell. "Children's Concepts of Average and Representativeness." *Journal for Research in Mathematics Education* 26 (January 1995): 20–39.

National Council of Teachers of Mathematics (NCTM). *Principles and Standards for School Mathematics.* Reston, Va.: NCTM, 2000.

Strauss, Sidney, and Efraim Bichler. "The Development of Children's Concepts of the Arithmetic Average." *Journal for Research in Mathematics Education* 19 (January 1988): 64–80.

13

Internet WebQuest: A Context for Mathematics Process Skills

Leah P. McCoy

O NE OF our most powerful technology tools is the Internet, which offers a wealth of easily accessible information. Mathematics teachers can effectively use this powerful resource as a tool in the teaching and learning of mathematics. One type of Internet activity, the WebQuest, is becoming increasingly popular in K–12 schools as teachers recognize its academic and motivational benefits for students.

WebQuest

Dodge (1995) defines a WebQuest as "an inquiry-oriented activity in which some or all of the information that learners interact with comes from resources on the Internet" (p. 10). Further, March (2003) says that a WebQuest is "an inquiry activity that presents student groups with a challenging task, provides access to an abundance of usually online resources and scaffolds the learning process to prompt higher order thinking."

The WebQuest format offers an environment that is ideal for implementing constructivist, learner-centered instructional practices (Dodge 2003, 2001; March 2004, 2000). The activity is typically a cooperative group project where students are presented with an interesting problem. They then seek data from Internet resources and use higher-level thinking to process these data. The final step is the preparation of a product that is an exposition or a defense of their conclusion. Scaffolding is furnished at appropriate levels by including links to useful Web sites, templates for the final product, and guidance on the organization of necessary tasks. Although WebQuests may involve many subjects as well as interdisciplinary activities, the focus in this article will be on mathematics WebQuests.

Editor's note: The CD accompanying this yearbook contains hyperlinked URLs for the Web sites that are relevant to this article.

Mathematics WebQuests

The *Principles and Standards for School Mathematics* (National Council of Teachers of Mathematics [NCTM] 2000) presents, along with the more familiar Content Standards, the five Process Standards: problem solving, reasoning and proof, communication, connections, and representation. All ten Standards, both content and process, should be integrated into mathematics teaching and learning at all levels. The WebQuest environment offers activities in which to apply these Process Standards in interesting and relevant contexts. See figure 13.1.

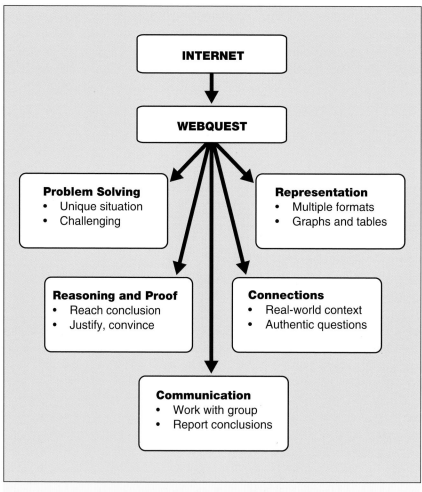

Fig. 13.1. WebQuests and process skills

Problem Solving

Problem solving is defined as the situation when students encounter a problem for which they do not know a solution method and must devise a method, solve the problem, and reflect on their solution. In Pyramid Puzzle (wcvt.com/~tiggr/index.html), students estimate the cost of building a great pyramid using modern materials and ancient methods. In teams, they plan, collect data, submit a bid for the job, create a scale drawing and model, and present a final report to the class. The WebQuest model allows the activity to be meaningful and realistic by actually locating and obtaining real data from Internet resources. To encourage motivation, topics of interest to students should be the focus of mathematical applications.

Reasoning

Reasoning is the process where students justify decisions and plan arguments based on their data. A typical WebQuest model involves answering a question by identifying needed data, finding the data on the Internet, organizing and representing the data, and preparing convincing arguments based on the data. A WebQuest that requires reasoning is Take Me Out to the Ball Game (warrensburg.k12.mo.us/webquest/baseball/). Students search the Web for hitting statistics (such as batting average, home runs, or runs batted in) to identify the greatest professional baseball hitter of all time. They use these data to identify and defend their choice of the best hitter of all time.

Communication

The *communication* of mathematical thinking is a multifaceted process, and practice can be facilitated by the WebQuest context. Many WebQuest activities are carried out in cooperative groups where students must work together to complete the task. This working together requires some discussion of the problem and the solution strategies. The simple act of putting thoughts into words improves the thought process. The product required at the end of the activity often requires students to prepare a presentation reporting or justifying their findings to the class or convincing the class of their point of view. The African-American Men and Women in Mathematics WebQuest (www.luxcasco.k12.wi.us/htdocs_oldsite/teacwebq/mathwebq.htm) involves selecting an African-American mathematician or scientist, conducting research on the Internet, and presenting this information to the class. Interesting variations include having students present this information as if it were a news report or as if they were nominating their mathematician for an award.

Connections

Connections between mathematics and contexts outside of mathematics are the heart of the applied problem of a WebQuest. The Internet provides contexts that are almost limitless, from weather data to nutrition information on fast foods to prices for comparative shopping. In Stock Car Racing (www.plainfield.k12.in.us/hschool/webq/webq114/index.htm), students form teams and design a car to participate in five main activities. They research racetracks, record track data in a spreadsheet, and decide which five races they will compete in. From a random drawing just prior to each race, they are assigned a position, and their points (and earnings) are assigned according to the NASCAR points awarded to the driver in that position in the real race. In another activity, they research Fortune 500 companies and select a company to approach as a sponsor for their car. They prepare a presentation to that company's executives to explain their budget and recruit the company as a sponsor for their car. They calculate all expenses and income on a spreadsheet. This is an extended activity, where students work on different aspects of the project over several months. Data are collected to make several decisions, and students must use the data in preparing an argument to recruit the sponsor. Students are able to see the relevance of mathematics in the authentic, real-world contexts, and they also see the connections among the different mathematical operations.

Representation

The final process skill is *representation*. Multiple representations are common in WebQuest problems as students organize and record the data they have collected, often using tables and graphs, and prepare convincing word arguments based on the data. Call Me (www.gowcsd.com/master/ghs/math/furman/linsystem/call_me.htm) is a WebQuest where students research several long-distance companies on the Internet and prepare a recommendation for which company would be best for their family. They represent the data in graphs and tables to highlight the differences among the companies.

Thus, all five process skills have extensive application within WebQuest activities. Whereas textbooks may be limiting and actual data collection may be logistically difficult, the Internet is a readily available data resource. Instructional activities based on the WebQuest model use this technology in developing and using skills in problem solving, reasoning, communication, connections, and representations.

An Example: ACC Basketball

In this WebQuest (www.wfu.edu/~mccoy/NCTM00/amy.html), ninth-grade algebra students worked in groups to explore statistics of basketball teams in the Atlantic Coast Conference. Students were assigned the task of discovering and explaining the relationship of offense (average points scored), defense (average points scored by opponents), and win-loss record. This WebQuest had three main activities, all using technology: (1) data collection from the Internet, (2) data analysis and representation using a spreadsheet, and (3) the presentation of results using a presentation program. At the end of the project, each student wrote an individual reflection.

Students were given links to the athletic departments of all nine ACC schools, where extensive data sets of statistics were available online. Although this information is also available in media guides and other publications, the Web offers a practical tool for students to locate and obtain the data conveniently and inexpensively. They were given flexibility to choose either men's or women's basketball and to choose any five teams. Nearly all groups chose to study all nine men's teams. Scaffolding was furnished in hints for locating the statistics on the athletic department Web pages, printed worksheets to help them identify the relevant data, and hints for calculating and presenting the needed statistics.

After locating the scoring statistics and recording them in a table, most groups represented the data in a double bar graph or a line graph showing the average points scored by each team and their opponents. See figure 13.2. They then based their conclusions on a visual examination of this graph. A typical conclusion was that "the winning teams had more scoring than the opponent, and the losing teams had even or less scoring than the opponent." Another group found the following: "There was a big difference in the scoring for the top three teams. It was closer for the middle three teams, and very close for the bottom three."

After some discussion, one group decided to calculate the difference between points scored and points allowed, and then graphed these data by team in order of rank. Their conclusion was that there was no clear relationship. This method brought up the topic of negative difference, since two teams scored less than their opponents. One student in the group suggested that they just change the order for those teams, and the differences would all be positive. Another student disagreed and said that order mattered, and if they had different orders, that would tell how far apart they were but not which was more. The teacher used this opportunity to explain magnitude and direction of differences. She also helped the group members to construct a bar graph with a y-axis scaled to less than zero. See figure 13.3. The members of

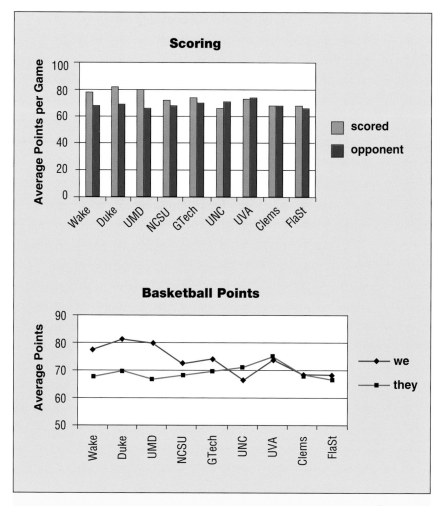

Fig. 13.2. ACC WebQuest: Bar graph and line graph

this group shared these mathematical concepts with the rest of the class during their presentation.

Another group had the most advanced plan. They used tables, graphs, and equations to compare the relationships between the winning percentage and both points scored and points given up. Having recently completed a unit on linear regression, these students used a spreadsheet to construct a scatterplot, give the trendline (regression equation), and give the R-square, which they used to find R (the correlation). See figure 13.4. Their conclusions were based on their comparison of the correlations.

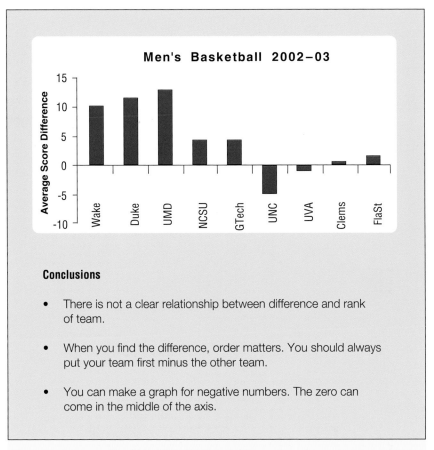

Men's Basketball 2002-03

Conclusions

- There is not a clear relationship between difference and rank of team.

- When you find the difference, order matters. You should always put your team first minus the other team.

- You can make a graph for negative numbers. The zero can come in the middle of the axis.

Fig. 13.3. ACC WebQuest: Differences slides

This WebQuest clearly involved all five process skills. *Problem solving* occurred throughout the activity as each group discussed their understanding of the problem, planned a solution, carried out the plan, and then reflected on their work. *Reasoning* was a clear focus as students analyzed and organized their data to justify their decision of whether offense or defense was more important. *Communication* was an important part of the activity in three ways. First, the group discussions were very rich, with considerable talk about both the basketball context and the embedded mathematics. The teacher monitored these interactions and became involved when groups needed direction or clarification. The second communication piece was the final product of the project, the presentation. This assignment required student groups to explain their work to the class in a computer slide-show presentation, thus compelling

What is the relationship between offense and defense and winning?

We found the points scored, the points given up, and the win-loss percentage. Then we made two scatterplots of the win-loss percentage and the points scores and of the win-loss percentage and the points given up. We had the spreadsheet also calculate the regression line and give us the correlation. *Y* is the winning percentage and the first *x* is points scored and the second *x* is points given up.

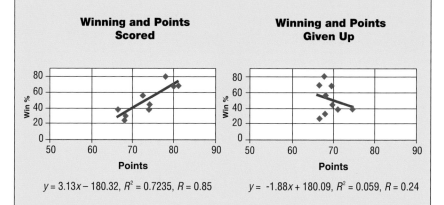

Winning and Points Scored

$y = 3.13x - 180.32$, $R^2 = 0.7235$, $R = 0.85$

Winning and Points Given Up

$y = -1.88x + 180.09$, $R^2 = 0.059$, $R = 0.24$

Conclusions

The line for winning and points scored is positive, meaning that as points go up, the winning percentage also goes up. The slope (rise/run) is 3.13, meaning that for every 1 point scored, the winning percentage goes up 3. The correlation is .85, which is a pretty strong relationship.

The line for winning and points given up is negative, meaning that as points go up, the winning percentage goes down. The slope is -1.88, meaning that for every point given up, the winning percentage goes down almost 2 points. The correlation is .24, which is not all that good.

We think that this shows that scoring points (offense) is more important to winning than how many points the other guys score (defense). If we had to draft a player, we would look for someone who could score.

Fig. 13.4. ACC WebQuest: Regression slides

them to further discuss and organize their collective thoughts. Third, each student individually reflected on the project and wrote a summary of his or her understandings.

The importance of *connections* was demonstrated first as a realistic context where the students were very engaged in a familiar and meaningful application. There were also connections in the mathematics, and some groups' methods involved diverse mathematical concepts. One group was familiar both with the construction of a bar graph and with negative numbers, and its members learned to express their negative differences in a bar graph. See figure 13.3. Another group chose linear regression, and in addition to comparing the regression lines visually, its members learned to assess and compare the correlation coefficients. See figure 13.4. All the groups used multiple *representations* of the data through tables, graphs, equations, and words.

Implications

Observations of mathematics students confirm the value of WebQuest activities. Students are actively involved, and a high degree of interest is evident. Because of the real-world aspect, students identify with their roles and have a real desire to understand and apply the appropriate mathematics. Teachers become true facilitators, offering advice and assistance as needed. Group discussion is goal oriented, since students share their thoughts and ideas. It is extremely interesting to hear students as they work together to draw conclusions from their data. They are doing mathematics; they are problem solving, reasoning, communicating, connecting, and representing.

WebQuests offer both breadth and depth in mathematics instructional activities. WebQuests can involve many different contexts and involve a wide variety of content and process skills. But at the same time, they are constructivist-based activities and require in-depth interaction between the students and the mathematics. Many WebQuest activities are available on the Internet, with topics as widely varied as vacation planning, rating roller coasters, and the stock market (Dodge 2003; March 2004; McCoy 2004). Teachers can easily follow these models to create additional activities that match the interests and level of their students. The integration of WebQuest projects throughout the mathematics curriculum motivates students and ensures the practice of process skills.

REFERENCES

Dodge, Bernie. "The WebQuest Page," November 8, 2003. (webquest.org), (June 18, 2004).

———. "FOCUS: Five Rules for Writing a Great WebQuest." *Learning & Leading with Technology* 28 (May 2001): 6–9, 58.

———. "WebQuests: A Technique for Internet-Based Learning." *Distance Educator* 1 (summer 1995): 10–13.

March, Tom. "Best Webquests.Com," January 8, 2004. (www.bestwebquests.com/), (June 18, 2004).

———. "Working the Web for Education: Theory and Practice on Integrating the Web for Learning," June 26, 2003. (www.ozline.com/learning/theory.html), (June 18, 2004).

———. "WebQuests 101." *Multimedia Schools* 7 (October 2000): 55–58.

McCoy, Leah P. "Math Webquests," January 5, 2004. (www.wfu.edu/~mccoy/NCTM00 /examples.html), (June 18, 2004).

National Council of Teachers of Mathematics (NCTM). *Principles and Standards for School Mathematics.* Reston, Va.: NCTM, 2000.

An Annotated List of Recommended WebQuests

ACC Basketball
www.wfu.edu/~mccoy/NCTM00/amy.html
Students collect and compare statistics for men's or women's teams. Students use graphs to convince their peers of the most important aspects. A good use of multiple representations of data.
GRADES 7–12
STATISTICS
GRAPHING

A Creative Encounter of the Numerical Kind
studenthome.nku.edu/~webquest/gabbard/
Students create a number system for aliens who have only four fingers. An excellent motivation in an interesting context.
GRADES 6–8
PLACE VALUE
NUMBER SYSTEMS

African-American Men and Women in Mathematics
www.luxcasco.k12.wi.us/htdocs_oldsite/teacwebq/mathwebq.htm
Student groups select a mathematician and present information to the class. A nice diversity element in the context of the history of mathematics.
GRADES 4–12
HISTORY OF MATHEMATICS

Call Me

www.gowcsd.com/master/ghs/math/furman/linsystem/call_me.htm

Students analyze different long-distance phone plans. Data are represented in graphs to support convincing arguments. A clear explanation of outcome product.

GRADE 9

MODELING

GRAPHING

REASONING

Conics in My Life

eprentice.sdsu.edu/J030J/ulloa-higuera/WQ/WQmain.htm

Students find real-world examples of conics and explain them. Then they create a new product using conics. An excellent connection of higher-level topics to the real world.

GRADES 9–12

CONIC SECTIONS

QUADRATIC EQUATIONS

Dilemma of the Dangerous Meat Loaf

imet.csus.edu/imet1/scotty/webquest/meatloaf.htm

Students analyze a meat loaf recipe and evaluate its nutrition.

A good connection to health with nice worksheets for scaffolding.

GRADE 7

PROBLEM SOLVING

REASONING

BASIC COMPUTATION

In the Time of the Old Ones

teacher.esuhsd.org/webquests/webquests/navajo/WQ102.html

Students create a Navajo rug pattern. A strong diversity context in Native American culture.

GRADES 3–5

GEOMETRIC PATTERNS

Lotto Fever

homepages.moeller.org/tfairbanks/lottofever.htm

Students analyze several state lottery games and design a lotto game. Promotes practical analysis using probability and requires the creative application of knowledge.

GRADE 6

PROBABILITY

REASONING

Math Models and Economics
www.amaisd.org/nheights/zennadi_&_james_project.htm
Students explore economic issues (car, home, budget, and investment). Comprehensive and well integrated.

GRADES 9–12
MODELING
REPRESENTATION

North Carolina Zoo
www.geocities.com/Athens/Parthenon/5827/breakout.html
Students plan a trip for zoo animals who have escaped. Excellent connections with science and social studies.

GRADES 4–8
MAP READING
BASIC COMPUTATION

Pay Now or Pay Later
campus.pc.edu/students/sbliff00/Secondary%20Math%20WebQuest
/index1.htm
Students compare data for CD prices and the number of illegal downloads. They use graphs to convince their peers of the relationship. Extensive Web resources are provided for understanding the topic.

GRADES 9–12
FUNCTIONS
GRAPHING
MODELING

Pyramid Puzzle
wcvt.com/%7Etiggr/index.html
Students calculate the cost of building a great pyramid today. The activity brings together modern-day construction and ancient wonders. Offers extensive use of mathematics.

GRADE 6
MODELING
GEOMETRY
BUDGETING

Stock Car Racing

www.plainfield.k12.in.us/hschool/webq/webq114/

Students analyze racetracks and compute points earned in races. They also prepare a business report to recruit a sponsor. Very engaging; has several different mathematical applications.

GRADES 7–12

BASIC COMPUTATION
GEOMETRY
PROBABILITY

Take Me Out to the Ball Game

warrensburg.k12.mo.us/webquest/baseball/

Students collect data on statistics related to hitting. Then, they select the greatest hitter of all time and convince others. Nice jigsaw cooperative group model.

GRADES 9–12

STATISTICS

Weather Forecast Showdown

www.uwm.edu/%7Ekahl/WebQuests/Showdown/

Students track weather forecasts and actual weather and rate the accuracy of the forecasts. Very clearly presented; includes a glossary of weather and mathematical terms.

GRADES 5–9

STATISTICS
REASONING

Weekend @ Bernie's

www.plainfield.k12.in.us/hschool/webq/webq103/index.htm

Students plan purchases for a new beach home. The activity offers a good variety in students' choices of items to research. The evaluation criteria are very clear.

GRADES 4–6

BASIC COMPUTATION
BUDGETING

Who the Heck Is Leibniz?

staff.jccc.net/jcrabtre/mission.html

Students determine the true father of calculus (Newton or Leibniz). Students are involved in real debate as they learn some history of mathematics.

GRADES 9–12

HISTORY OF MATHEMATICS

14

A Vehicle for Mathematics Lessons: In-Service Teachers Learning to Use PDAs in Their Classrooms

Julie Cwikla

We get handed (technology) stuff all the time that we're never given any time to develop.

—A middle school mathematics teacher

The real economic power of invention is realized only when innovations are widely adopted by the people who can benefit from them.

—Lewis Perelman, *School's Out*

TYPICALLY, teachers in grades K–12 are introduced to new technological innovations in a workshop type of environment. These workshops often consist of traditional training about the ins and outs of spreadsheets, slideshow presentations, Web design, and so on, without any connection to teachers' classroom practice. Teachers might learn how to use a new software product, but without specific conversations about the benefits to students' learning beyond teachers' existing practice, minus the trade-off for time-intensive start-up and development, teachers often enter their classroom the next morning little changed. Engaging teachers in meaningful dialogue about technology and the ways they envision technology in their daily practice helps teachers connect innovations to their mathematics lessons. Otherwise, technology is viewed as a curriculum "add on" rather than as a powerful "vehicle" for mathematics learning and investigation. This article will review the progress of the Teacher Research Team, a group of middle school mathematics teachers on the Gulf Coast in Mississippi collaborating to design lessons that use the powerful technology of personal digital assistants (PDAs). The purpose of this arti-

Editor's note: The CD accompanying this yearbook contains an Excel file and a PowerPoint presentation that are relevant to this article.

cle is to document the teachers' development, how they made sense of the technology, and the issues and difficulties that arose in our discussions.

In the National Council of Teachers of Mathematics (NCTM) 2000 Yearbook, *Learning Mathematics for a New Century,* Waits and Demana (2000) predicted that "the marriage of calculators and computers will allow us to resolve some of the intractable equity issues of our educational system" (p. 64). That union has occurred in personal handheld computing devices or personal digital assistants. PDAs have been used for years in the professional workplace, and these inexpensive tools have large-scale implications for a changing learning environment for K–12 mathematics students. The same might have been said for desktop computers in the late 1980s as they entered the schools. Nevertheless, desktop computers have *not* become an integral part of the K–12 mathematics classroom, and lessons do not look much different from the way they did twenty years ago (Jacobsen, Clifford, and Friesen 2002). Moreover, school computing technologies average $7 billion a year, yet students' computer access is measured in minutes per week (Soloway et al. 2001).

In a review of past, present, and future applications of calculators in the mathematics classroom, Waits and Demana (2000) claimed that "the lesson we learned is that *change can occur if we put the potential for change in the hands of everyone"* (p. 53). I predict that the same lesson will be learned as handheld devices become more widely used by students and teachers in the K–12 classroom. Desktop computers were not accessible to every student at all times because of cost and sheer mobility, whereas PDAs offer sleek, affordable convenience in every learner's hand and offer the same capabilities of a laptop in the 1990s that cost $4000 (Swan et al. 2002). Handheld devices allow schools to offer the ubiquitous computing capabilities similar to that which professionals experience in the workplace.

Ostlet (2002) termed the devices the "21st century Swiss army knife" of computing. Because of their size and capabilities, PDAs can be used in mathematics classrooms in ways not even approachable by previous technology. Personal handheld computing devices will alter the mathematics classroom as well as other subjects. Software exists that allows teachers to "beam" quizzes to students; they are beamed back when completed, graded, and entered into an electronic grade spreadsheet. In addition to easing the teacher's role, the strength of a computer in every student's palm will furnish students complete access to word processing, spreadsheets and charts, presentation software, graphing calculator capabilities, a calendar for assignments, e-mail, and other academic software packages and organizers. Students can also beam information to each other to share findings in a group investigation. In addition, at the writing of this article advances are being made to merge graphing calculators and laptop computer capabilities. Furthermore, the TI Voyage 200 has some PDA-like capabilities built in along with the TI-92 Plus capabilities. However, regardless of the type of technology used and as documented with other technological advances and attempted K–12 classroom integration, if teachers are not properly prepared *and* supported to learn about the innovation, they will not have the opportunity to design daily lessons that use the technology.

When technology is presented and used as a "vehicle" for learning and not an "add on" to the already immense list of frameworks and topics U.S. teachers need to address each year, teachers are more likely to shift from their view of technology as a separate entity. This article presents an overview about how the Teacher Research Team (TRT), a group of four middle school mathematics teachers, began technology infusion one classroom and one lesson at a time. Teachers were given the time and opportunity to work and talk through their own fears about technology use and develop a lesson for middle school students. This article describes the teachers' dialogue and their path to lesson development.

The TRT Context

The Mississippi Gulf Coast is an ethnically and economically diverse area of the United States. The ethnic makeup of the population is continuously changing with the booming travel and tourism surge, along with the active seafood industry. Across the coast the population consists of three major groups: African Americans, Asians (many of whom are learning English as a second language), and Euro-Americans. Many families in the state of Mississippi and on the Gulf Coast are economically disadvantaged. Several schools have more than half of their students receiving free or reduced lunch, and although by the year 2001 Mississippi was able to place at least six desktop computers in every classroom (Mississippi Department of Education 2001) and offered technology workshops, teachers' practice and everyday lessons were not integrating technology and mathematics content effectively, if at all.

It was important for the TRT to begin immediately thinking about technology and the PDAs as potentially part of their daily practice and not an "add on." These tiny devices have extraordinary potential to change the learning environment for K–12 students. However, without much guidance in the literature about using PDAs in the middle schools at the time of this study, a group of professionals was assembled to develop, pilot, video record, and collaboratively analyze mathematics lessons using the PDAs. These teachers were knowledgeable, motivated, and dedicated. Each mathematics teacher had been recommended by university colleagues as well as by their school principals. The four teachers, two female and two male, were each from different schools, with the thought that they might be seeds for technology at their building in the future and that being from different schools might avoid any possible building-level politics. My role was as participant, observer, and technology consultant, learning about the process of teacher collaboration and mathematics and technology lesson development. Using the Japanese lesson study model, which is grounded in classroom practice and stems from the teachers' experiences and expertise (Lewis 2002; Stigler and Hiebert 1999), these teachers would carefully develop a lesson that used the PDA technology with special attention given to their students' mathematical strategies and understanding.

Engaging Teachers in Dialogue

The first meeting consisted of introductions and a sharing of experiences followed by (1) an overview of Lesson Study, (2) a discussion of the TIMSS video analyses and the types of student-centered lessons desired, and (3) the possible goals of this group, later self-named the Teacher Research Team. The philosophy that I shared as a teacher educator stressed that my purpose was to help the teachers become researchers of their own practice, provide technology equipment, and help them access the technology, retrieve useful resources, and help them improve their practice one lesson at a time. However, the specific goals and direction of the group were to be determined by the teacher-researchers.

The teachers were given handheld devices manufactured by Palm loaded with the Documents-to-Go 5.0 software, which included software similar to Microsoft's Word, Excel, and PowerPoint. Several other handheld devices are also available that have a Microsoft Windows–based operating system and have a screen appearance more similar to a desktop computer. The teachers explored the equipment, learned to use the stylus, entered data, and navigated the Palm operating system. The teachers were presented with readings about Japanese Lesson Study from Lewis (2002) and Stigler and Hiebert (1999) to review if desired for the next meeting. The first meeting lasted a little more than three hours, and the group set the date for its next meeting.

By the second meeting two weeks later, the teachers were comfortable with the equipment, using the PDAs for personal data with the address book, to-do list, and calendar. The teachers were also visibly more comfortable with the setting, a university computer lab, and with me, a university professor, and my goals and intentions. To understand more fully the TRT teachers' dialogue about technology, it is important to understand their previous experiences in technology workshops and training in their individual schools and districts. The TRT had the following exchange about PowerPoint during our second meeting:

> *Rose:* A year and a half ago we were all trained on PowerPoint. Two days with PowerPoint. Have I ever sat down and made a PowerPoint presentation to use in my classroom? No!
>
> *Bob:* Same here. Same here.
>
> *Anna:* Because, for one thing, you have to figure out where you would plug it in, and is it a good use of your time? Is it worth the use of your time [in developing] to show them that?

Rose and Anna are both Nationally Board Certified Teachers. Rose is in her mid-fifties and Anna is in her late thirties. Bob is a retired military officer, and teaching is his second career. He is in his late fifties. Tom, the fourth teacher on the team, is introduced later. Pseudonyms are used for the teachers of the TRT.

Besides the presentation software, the teachers also did not feel comfortable with the use of spreadsheets, even though they had attended workshop training sessions, which are part of the middle school standards in Mississippi.

> *Rose:* Another confession: I've never used Excel. Never.
>
> *Bob:* I haven't either.
>
> *Anna:* It's easy. And I can't do much with computers. But I can do that. But you can put a formula in.… Because I went to the workshop where you had to do your computer, your grade book on there, which I never did, but I did figure out, you know, that you could plug stuff, that you could put a formula in there, and tell it, you know, this column and this column.
>
> *Rose:* Right. I went to a workshop a long time ago where they said, "Yeah, you can do this." And we did a few. And that's the last I ever touched it.

The teachers' experiences with software and technology up to the spring of 2003 at the start of this project were far removed from their daily teaching. They had received traditional workshop training, but it had not resulted in their using the software in any teaching or learning capacity.

However, it was not true that the teachers did not *want* to use technology in their practice. But they had not been given the opportunity to plan lessons that incorporated what they had learned in the workshops, and it was not clear that they needed to use technology to improve their current practice. Rose commented that it is not enough to be given technology and software training, "You have to *need* to do it. You don't just sit down and say I'm going to learn to do this today. You have to *need* it, that's the whole thing." The TRT created a need for these teachers to learn about the technology, and the group goal of creating a PDA lesson for their students forced them to sit down and take the time to do it. Similarly, Anna defended teachers by saying, "I think most teachers would like to do some of the things that they're taught to do, but you have this textbook that tells you exactly what order to go in, and that's what most people are going to do. And figuring out how to plug it in is really hard." Inherent in Anna's comment is her belief that technology is an "add on" because it must be "plugged in" somewhere as opposed to serving as a vehicle for exploring, recording, and analyzing the mathematical content built on the confines of the textbook. This mind-set was prevalent in the group and is further addressed in the next section.

During our first two meetings, which totaled eight hours of collaboration, the teachers shared and at times vehemently expressed their apprehensions about technology and their reservations about finding time to *add* something else to their curriculum. Some of their apprehensions were due to their own experiences with technology and technology training in the past, but other concerns were related to more general systemic restraints, such as (1) finding time to fit technology lessons into the existing curriculum, (2) the logistics of the new equipment and allowing time to teach students about the PDAs, and (3) weighing the benefits of technology over their existing practice.

"What Do I Take Out If I Add Technology?"

The teachers discussed systemic restrictions that they had little or no control over, such as the existing curriculum, state frameworks, accountability mandates, and state assessments. The teachers all seemed quite overwhelmed with the notion of adding something else to their plate. Anna shared the following comment with intensity:

> I just wish they (the state) would prioritize what they want us to do. So that we could say, you know, "Of these 60 things, how much time would you like me to spend on each one, and I'll do it? Just tell me. What's the most important thing to you, so that we can do what you really want us to?"

Tom, the fourth member of the group, is a young first-year teacher in his mid-twenties. He related the Japanese Lesson Study readings to his classroom:

> I guess the U.S. opinion is that more is better. You know, the more you try to cover in one year, you know, the better off we're going to be..... Like I

said, you can't present all this stuff, like she was saying, and then have time to maybe do an activity that would make them, make that light go off, and go, "Oh, you know, well, now I know when I would do that." Or, "Now I know, I see when I … " I mean, you know, you can't do that.

Each of the teachers struggled with the "quality versus quantity" content issue, and it continued to resurface in the TRT discussions periodically, depending on events such as possible pending changes in state or district policies or a conversation at a faculty meeting. Anna came to one of the TRT meetings following a school faculty meeting where teachers were being asked to do more writing across the curriculum. "You can't lose days," she said. "You know, sometimes they (administrators) say things like, 'Well, you can fit this in.' Like you were, like you had all stopped to have a coffee break, or something!" Time was a persistent issue: (1) time for students to learn to use the PDAs, (2) time to use the PDAs during a lesson, and (3) time for the teacher to evaluate their use.

> *Anna:* But see, to teach the section in the book that's on perimeter and area can take a day and a half, and to do those (PDAs), and evaluate them well, will take five days. And so what you do is, you say, "I don't have time, I've got to get to the next."
>
> *Tom:* Uh huh (*nodding*).
>
> *Anna:* You know, it's really hard, too…. Because it's, we would like it to be about understanding, but so much of it is about *volume*. It's really hard not to feel pressured in that way.
>
> *Rose:* I think the first go around, I feel hesitant to even have five (PDAs), and hand them out to the class. I'm not … because the time…. Since I'm still new at this (the PDA), I'd rather do what I'm going to do, and that's it.

The teachers considered (1) using their own PDA with a projector in the front of their classroom as an activity for the whole class, (2) providing one PDA to each group of students, and (3) furnishing every student his or her own PDA to hold and use.

> *Bob:* Well, I don't know. I don't know. It would be fun if we could do it with the kids having it.
>
> *Rose:* But justifying the time spent on that is the catchy part. Because, you know, the bottom line is what *learning* are they going to have as a result of this?

After a lengthy discussion, the teachers decided to furnish PDAs to all their students or to pairs of students, depending on the number of students in their classroom. However, they still wrestled with determining how and if using the technology was going to strengthen a lesson and which lesson to choose. While brainstorming and debating the worthiness of different topics, the teachers reflected on their current practice and asked themselves questions such as: "Will this be an improvement?" "Is this worth the time involved?" "Will the students be more interested and will they develop a deeper understanding?" An example of one such exchange follows.

> *Anna:* But would this be any better than … I mean, it seems like if you're going to do it on the computer, that the computer's better, you know. Because it's, for one thing, it's bigger, and they already know how to use it. How would this be better than using a computer, or would it?
>
> *Tom:* You're talking costwise.
>
> *Anna:* Well, I'm just talking general-wise. At least they know how to [use the desktop]. They're used to that, and it would be familiar to them. For what we're talking about, would this [PDA] be any better to do, like the Excel program thing, than a computer, or would it?

The TRT later agreed that the PDAs would encourage more flexible student collaboration and conversation, allowing the students to sit at their desks and in groups facing one another. Then the dialogue shifted to issues of logistics and how to teach the students to use the device.

"Hold On, Class, Push the *Other* Button!"

During the first three meetings and even at times up until the teachers enacted their lesson, they continually addressed their concerns about how the students would react to the equipment. This concern permeated the group's beginning discussions. The teachers talked about several apprehensions: (1) the time required to teach students how to use the equipment, (2) fears about damaging the equipment and using the stylus as a weapon, (3) students opening other software and not staying on task, and (4) keeping all the students in more or less the same place. Examples of these concerns are illustrated in the following excerpts. The first exchange is from the second TRT meeting, which was an afternoon meeting that lasted just under four hours. In the beginning, the teachers fed off one another's fears, but this built trust and camaraderie over the course of the meetings. At times there was a sense that the teachers might not use the equipment at all.

> *Bob:*   If we're going to create a lesson where the students are hold-
> ing one of these handheld computers, we're going to have to
> spend a good part of that time introducing them to it, so
> they'll know how to use it. And that would probably take up
> an entire lesson, I would think.
>
> *Tom:*   Yeah, um-hm.
>
> *Bob:*   Depending on what we're teaching. I still don't have a real
> clue as to how, you know, we're going to step through this.
> But if I were a student getting this, and I was told to solve
> this math problem, I'd need to know a lot about this.
>
> *Anna:*   And you'd want to play with it first.

Tom, a younger teacher, was afraid that students might damage the equipment. "I mean, the way that they destroy stuff. You know. Not knowing how
to do it and they … I mean, just the maintenance on these things." And Bob
referred to the stylus, adding in a somewhat joking manner, yet still concerned, "And this could be a weapon." The teachers were also afraid of students on the equipment wandering away from the spreadsheet tasks and
"playing around" with computer games, graffiti, other software, and tools
such as the note pad. Resigned to the fact that the students would need to play,
they discussed when would be the best time to let students "play" and explore
on the device.

> *Rose:*   I'm almost inclined to doing the playing at the end, when
> they complete the task.
>
> *Anna:*   Oh, that won't happen. I mean, I just think mine are going
> to play whether I let them play as soon as they get them or
> not. They're going to play as soon as they get them.

Bob disagreed with Rose and wanted to encourage students' curiosity about
the equipment and its capabilities. "I think they could play with them in the
beginning…. Curiosity, I don't want to deprive them of it. They're going to
be curious." Later he offered a way to guide students' curiosity and help them
learn how to use the device while exploring.

> *Bob:*   One of the things I was thinking about when you were talk-
> ing about [students' distraction] and what not, is maybe let
> them play for a few minutes, and then direct them to a
> PowerPoint activity that they work through. There's an
> introduction to what you're going to do. So they can flip
> through some slides to get them thinking about the topic,

and then put the handhelds down, then you do your lesson, then they come back to the Palms.

In addition to the issue of keeping the students on the correct handheld task, Rose was worried about the students accidentally removing the programs from the equipment.

Tom: I'm worried about them erasing what we put in there to do.

Anna: Oh, they're not going to do that. Because they're not going to go to the spreadsheet, they're not going to care about documents to go. They're going to want to see, play with the calculator, or something that they can find on there. They're not going to find our documents.

Much of the conversation up to this point was about how to control the students from doing "this and that" with the equipment. The third TRT meeting was a daylong professional session, and the level of dialogue made an explicit shift. Instead of focusing on the "don'ts" for their students, they began problem solving as a group and decided what could be done so that they could minimize distraction within reason and get the students interested in the mathematics. Rose and Anna had the following exchange; there is a shift here for Anna, who earlier was afraid of her students being off task during the entire lesson, playing games.

Rose: I ran an elementary school computer lab for about eight years. Kindergarten through sixth grade. And I know that when I was teaching, they each had their own computer. When I was teaching a lesson, some people listened, and heard, and were at the right place at the right time. And the rest of the people got lost, or got behind, or hit the wrong button, and then there was an inordinate amount of time making sure everybody was [on track]. And so I have a little bit of fear that with this, that the more complex it is, the more time we're going to lose, in "Oh, you're not on there? Hold on, class."

Anna: And it's valid. I think they will. But that will be the nice thing about doing the PowerPoint [whole group, in front of the class], which, frankly, scares me. But I think if we have, okay, "Step one, up here." And it would be nice even if we just had a picture of the [PDA] screen, and the little finger pushing the button we want them to push.

This exchange gives an example of how the teachers in the TRT validated one another and helped others on the team through some of their concerns, as opposed to an "expert" facilitator offering advice and directions, which at times are not taken as seriously by teachers.

The teachers were becoming more comfortable with the idea of equipment in the hands of every student, and they began to work through some of the specific logistics and methods to lead students through becoming familiar with the equipment. It was clear they were all on board and ready to tackle this as a team. The teachers worked together to determine how the PDAs would be able to enhance their current mathematics lessons, beyond simply the intrinsic value of having a new piece of technology in the classroom.

"How Do I Connect This Device with a Math Topic?"

From the first meeting, teachers had a difficult time imagining how the PDAs would connect with their current practice and how it would help their students learn. As the teachers experimented with the technology on their own and began talking about how the devices could be used in their classrooms, the group reached a consensus about the PDA capabilities. However, there were comments and moments of confusion, as exhibited in the following discussion from the second meeting:

> *Rose:* I'm just not sure about connecting this piece of equipment with the topic. I just don't really see the connection yet. Am I the only one?
>
> *Bob:* No. I feel that way.
>
> *Anna:* Well, you could use it, like…. I can see how you could use it if you had a formula, and you wanted to see what happens if you changed a variable. Or if you made this variable bigger, you could do inverse variation with it.

Part of the teachers' block in imagining the uses of the PDA and Documents-to-Go was their own limited use with computers in general. Most of them had never used spreadsheets or presentation software outside of a workshop or two. Not knowing the powerful capabilities of the Documents-to-Go software, which included word processing, spreadsheets, graphing capabilities, the power to create and display presentations, and numerous other programs for the Palm operating system available for trial on the Internet, created barriers to envisioning uses of this equipment in their own classroom. Anna openly shared her confusion, "How do you know what the software can do? Because I looked at a little bit of that, but I didn't really understand exactly.…

I want to know what kind of lesson plan this would go with?" This piece of equipment did not fit in a ready-made lesson plan structure that the teachers could imagine. Likewise, at first Rose could envision a possible mathematics lesson, but one that would be more of a low-level demonstration and not the critical thinking and engagement prevalent in a *Standards*-based classroom that is typical of her practice. The group dialogue had yet to produce a vision of an effective mathematics lesson using this new technology.

> *Rose:* Um … this is going to sound horrible, but I have in the back of my mind that I'm to develop something to use on that [PDA]. I'd really rather find something like he was talking about, the tables for the percent things. And I don't do a lot of demonstrating, but what I'm seeing is, putting this [slideshow and spreadsheet] up as a demonstration as part of the whole-class discussion.

> *Anna:* Yeah. But if they each had one in their hand and you could say, "Okay, push the little A, B, C, thing and type this formula in." And they could do that, and then you could say, "Okay, let's make …" And surely the software is there that you could say, "If A is this, and B is this, and C is this, put enter." I don't know how you do that, but, you know, and "What do you get? Now let's change this. Now, what does this stand for again? Oh, that's your interest rate. Well, what would happen if we made that interest rate 5 percent? First, everybody guess what they think would …" I mean, you could do that. That would be a useful thing.

Anna, although like the others, lacking in technology expertise, was a catalyst for the group in brainstorming lesson ideas and verbally sketching ways teachers could connect their ideas to the possible uses of the PDA. However, the group, as well as each individual, oscillated between excitement and fear. When they did visualize their classroom with a room full of their middle school students, each holding a relatively expensive piece of equipment, it was difficult for them to see the vast opportunities over their reservations illustrated above. But eventually the teachers made significant headway on the actual lesson development.

"What Should Our Lesson Look Like?"

The teachers came up with several topics for the lesson they would design using the PDAs. The teacher exchange and brainstorming over three meetings is testament to the multiple applications of the PDA in the middle school mathematics classroom. Here are a few examples with which the group wrestled and that illustrate the diversity and expertise of these teachers' ideas and the power of interactive graphs, pictures, and spreadsheets in each student's hand:

Bob: If we could agree on a good geometry lesson, somehow showing the dimensions, you know, with [the PDA], this is going to look so much nicer than anything I can put on the board. So if there's some way we can use this to show them something that really looks slick, and it's going to be eye-catching and everything, I'd like it just for that. You know. Just for the motivation. Build up the intrinsic motivation to look at the thing.

Anna: Maybe if they had something on there where they could see the correlation between what percentage is taxed or what percentage they borrow their money at. You could give them each … choices. "You want to borrow $10,000 at 6 percent or do you want to borrow $20,000 at 3 percent, and which one over five years is going to end up costing you more?" … They could just manipulate it. And percent seems to be one of those areas that are on all their testing, [and] they have a hard time [with it].

Tom: I guess this was more basic, what I was thinking. You could do the fractions and the decimals. Almost look like on a slide show. Have it to where you could see, you know, how they were equal. You know, 2/3 is equal to 4/6. You know, have it overlap, you know, through the progression. And see how the fractions turn into decimals, stuff like that.

Agreement on a lesson topic was reached after more than fifteen hours of meeting time, and it was not because the teachers were in opposition. But finding a topic that was meaningful to all of them, that was a good fit in their academic sequence, and that they could all envision using with the PDAs required heavy discussion about learning and teaching. The TRT decided to combine data analysis and visual representations with pie graphs created using spreadsheet software. The lesson format was structured using presentation software in front of the whole class and was presented to each student on the PDA. The lesson is summarized in the lesson outline below. Details about the lesson along with classroom vignettes can be found in Cwikla and Morse (2004).

Lesson Outline: "Who Wants Pie?"

1. The teacher led a whole-class refresher of fractions, decimals, percents, and degrees in a circle using a SlideshowToGo presentation, analogous to the PowerPoint presentation WhoWantsPie.ppt on the CD accompanying this yearbook.

2. Student pairs sorted the twelve months of the year by the number of syllables in each month's name into a frequency table (fig. 14.1).

Who Wants Pie?
Making pie charts with a handheld computer

1. Calendar Frequency Table

4 Syllables	3 Syllables	< 3 Syllables

2. Calendar Spreadsheet

Category	Frequency	Fraction	Decimal	Percent	Degree

Fig. 14.1

3. Then they completed the frequency table to convert the syllables frequency to fractions, decimals, percents, and degrees in a circle.

4. Next, they turned on their PDA and opened up the SheetToGo software to the file named "Syllables" (see fig. 14.2), with rows and columns already labeled to correspond with the frequency table in figure 14.1. The students were given a sheet of directions on opening the SheetToGo, on data entry, and on plotting.

◆	A	B	C	D	E	F	G	H	I	J	K	L
1	Category		Amt. in category		As a fraction		As a decimal		As a percent		In degrees	
2	4 syllables		2		1/6		0.1667		16.67%		60	
3	3 syllables		4		1/3		0.3333		33.33%		120	
4	less than 3		6		1/2		0.5		50.00%		180	
5												
6												
7			12									
8												

Fig. 14.2

5. Each student entered numbers in the table and created a pie chart to display the relationships. Students checked their PDA pie-chart displays with one another and also with the degrees computed in their pencil-and-paper spreadsheet. The teacher then displayed a pie chart with the SlideshowToGo presentation so that the students could verify their own PDA display. See figure 14.3.

Fig. 14.3. Pie chart displayed for class as teacher assists and checks individual students.

6. After this somewhat directed minilesson, students continued to work in pairs to determine how they spend the twenty-four hours of a day (e.g., 7 hours in school, 2 hours of sports, 1.5 hours of homework). Then, as with the syllables task, they found the fraction of a twenty-four-hour day spent doing each activity, entered the data into a new, clear spreadsheet where they labeled the categories, and created another pie chart on their PDA.

7. Pie charts were compared with peers and a whole-class discussion of the number of degrees and hours in the day spent doing each activity ended the approximately fifty-minute lesson.

Lessons Learned

The middle school students enjoyed using the PDAs, which enabled each student to use a spreadsheet. Because of the lack of computer facilities and the difficulty in scheduling computer lab time, few of these students had ever used a spreadsheet to compute percentages or to plot data. The PDAs allowed individual access for every learner, and although the teachers were concerned about the students' treatment of the devices, few problems were encountered. In addition, because students were talking and helping one another to use the technology, this activity led into discussing their resulting pie charts. Even though the equipment might tend to lead students toward individual, self-sufficient mathematical investigation, there was significant sharing, comparison, and mathematical conversations among the students about the hours spent in the day playing sports, the percent of a day, and the corresponding slice of their pie chart. As each teacher implemented the technology research lesson in their classroom, their classroom was video recorded.

Following each implementation, the TRT met to view and critique the lesson before the next teacher's implementation. Changes to the lesson were as follows: (1) The teachers decided to number each of the PDAs and assign a number to each student in order to create a sense of accountability for the equipment. (2) With respect to the lesson, one teacher decided to add her own introduction to the lesson, asking students to tally as a whole group their preference for desserts of different temperatures as an introduction to frequency. (3) There were minor wording changes made to the lab sheets distributed. (4) As the lesson moved from teacher 1 to teacher 4, the whole-class SlideshowToGo became more abbreviated and less teacher-directed as a result of their watching and discussing the video.

One of the teachers commented that on reflection she thought the lesson was too teacher-centered. Although we agree, the purpose of this project was to allow the teachers to develop a task and method meaningful to them and their curricula. However, it is not surprising, given the teachers' initial fear and intimidation about technology, that the lesson was developed in a more traditional and controlled manner. This was only their first PDA lesson, which allowed the teachers to overcome their reservations about using handheld devices in their classroom.

Implications

The development of the TRT was designed to incorporate research-supported professional development practices (see Cwikla 2004). The TRT was a small group of teachers who were committed to the process and practice-centered philosophy of this project as well as to providing technology experiences to their students. Because of the support and camaraderie they found within this group, they made time to collaborate and develop technology-motivated lessons after school. These teachers and their dialogue serve as a guide for how classroom professionals wrestle with issues as they begin to learn about a new piece of technology and the ways to promote students' understanding of mathematics and enhance their daily practice. Their dialogue and struggles offer a model for other teachers working to use new technological advances as well as an example for professional developers working to help teachers advance their practice and prepare their students for the workplace of the future.

REFERENCES

Cwikla, Julie. "Show Me the Evidence! Mathematics Professional Development for Elementary Teachers." *Teaching Children Mathematics* 10 (February 2004): 321–26.

Cwikla, Julie, and Timothy Morse. "Handheld Computers: A Teacher Research Team Develops Technology Driven Lessons." Paper presented at the Annual Meeting of the National Council of Teachers of Mathematics, Philadelphia, Pa., April 2004.

Jacobsen, Michelle, Pat Clifford, and Sharon Friessen. "Preparing Teachers for Technology Integration: Creating a Culture of Inquiry in the Context of Use." *Contemporary Issues in Technology and Teacher Education* [Online serial] (March 2002). Available at www.citejournal.org/vol2/iss3/currentpractice/article2.cfm.

Lewis, Catherine. *Lesson Study: A Handbook for Teacher-Led Instructional Change.* Philadelphia: Research for Better Schools, 2002.

Mississippi Department of Education. *Process and Performance Review Guide.* Jackson, Miss:, Mississippi Department of Education, 2001.

Ostlet, Elliot. "PDA's: The Swiss Army Knife of Handheld Technology for Mathematics Classrooms." *Society for Information Technology and Teacher Education International Conference* 1 (January 2002): 1102–5.

Soloway, Elliot, Cathleen Norris, Phyllis Blumenfeld, Barry Fishman, Joseph Krajcik, and Ronald Marx. "Log On Education: Handheld Devices Are Ready-at-Hand." *Communications of the ACM* 44 (June 2001): 15–20.

Stigler, James, and James Hiebert. *The Teaching Gap: Best Ideas from the World's Teachers for Improving Education in the Classroom.* New York: The Free Press, 1999.

Swan, Kathleen O., Gerry M. Swan, Stephanie D. Van Hover, and Randy L. Bell. "A Novice's Guide to Handheld Computing." *Learning and Leading with Technology* 29 (May 2002): 22–27.

Waits, Bert, and Franklin Demana. "Calculators in Mathematics Teaching and Learning: Past, Present, and Future." In *Learning Mathematics for a New Century,* 2000 Yearbook of the National Council of Teachers of Mathematics (NCTM), edited by Maurice Burke, pp. 51–66. Reston, Va.: NCTM, 2000.

15

Using the Internet to Illuminate NCTM's *Principles and Standards for School Mathematics*

Eric W. Hart

Sabrina Keller

W. Gary Martin

Carol Midgett

S. Thomas Gorski

THE National Council of Teachers of Mathematics (NCTM) Illuminations project was launched in 1998 to support NCTM's *Principles and Standards for School Mathematics,* which was then still under development. The mission of Illuminations is to improve the teaching and learning of mathematics for all students by providing high-quality Internet resources that "illuminate" *Principles and Standards*.

Overview of Illuminations

The first author of this article became the director of the project in 1999, and the other authors joined shortly thereafter, forming the development team that conceptualized the project and built the Illuminations Web site (see fig. 15.1). From 1999 until 2003, when our participation on the project ended, the Illuminations Web site grew to become one of the leading school mathematics Web sites on the Internet. Visits to the site increased from only a few hundred in September 1999 to about 331,000 visits in October 2003. Each visit is one person staying on the site for an average of about eight minutes; the number of "hits" (clicks on any link on the site) in October 2003 was more than 9.75 million (WebTrends report, October 2003). Note that since this article was written, the site and its content have been redesigned and will continue to evolve, so that some of the details discussed in this paper may be different from what is on the site today.

Editor's note: The CD accompanying this yearbook contains an electronic version of this article, complete with hyperlinked URLs.

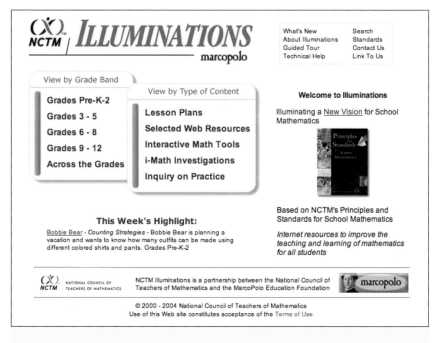

Fig. 15.1. Illuminations home page in January 2004

Illuminations is a partner in the MarcoPolo Internet Content for the Classroom program (www.marcopolo-education.org), funded by the MarcoPolo Education Foundation. The partners in this program are NCTM, the American Association for the Advancement of Science, the National Council of Teachers of English, the International Reading Association, the National Council on Economic Education, the National Endowment for the Humanities, the Kennedy Center for the Performing Arts, and the National Geographic Society. Additional funding for the project has been generously provided by the GE Fund, the philanthropic foundation of the General Electric Company.

The mission of Illuminations has been carried out by pursuing four main development strands:

1. Online, interactive, multimedia resources (primarily using applets and videos)

2. Internet-based lesson plans

3. Reviewed and categorized external Web resources

4. A Web design framework that organizes and presents the content in such a way that the design itself helps illuminate *Principles and Standards* and makes all content as usable and accessible as possible

The Internet resources on Illuminations serve many purposes for many audiences, by furnishing professional development and teaching resources for teachers, offering rich classroom materials for students, communicating the vision of *Principles and Standards* for all users, and providing outreach and an Internet portal for NCTM. Primarily, though, these resources are for teachers, to help them understand and enact the vision of *Principles and Standards* and to teach mathematics more effectively to their students.

In this article we will discuss each of the four development goals listed above, and we will present informal case studies of how Illuminations has been used for professional development with preservice and in-service teachers. But first we discuss some general principles that guide the development work.

Guiding Principles for Site Development

The overarching principle guiding the design and content development is the basic goal to illuminate *Principles and Standards.* Moreover, other guiding principles are drawn directly from this document. For example, we have developed the Web site on the basis of the recommendation that "[s]tudents' understanding of mathematical ideas can be built throughout their school years if they actively engage in tasks and experiences designed to deepen and connect their knowledge" (NCTM 2000, p. 21). Likewise, to truly understand the vision put forth in *Principles and Standards,* one must actively engage in doing mathematics in a manner that is reflective of that vision. Thus, we have focused on creating peer-reviewed resources that engage teachers, students, and all visitors to the site in doing and thinking about mathematics.

As all teachers know, a principal factor in successful teaching and learning is student motivation. Related to this idea, a research study in the *Journal of Computers in Mathematics and Science Teaching* states, "In summary, we observed that the motivational aspect of learning with the Web likely results from a number of factors, including ... interactivity, variable entry, data base fascination, and multimedia" (Moor and Zazkis 2000, p. 101). We have developed Illuminations with these factors in mind. For example, interactivity and variable entries are built into the Illuminations interactive mathematics investigations, database fascination is exploited in the carefully organized collection of reviewed external Web resources, and multimedia experiences are prevalent throughout the site.

Central to the development of the Illuminations resources is our conceptualization of the Internet as a global, online, interactive, interconnected, multimedia platform on which one can build and deliver powerful educational and professional development experiences. How can effective Web-based learning environments be built on this platform? An article in the *Journal of*

Research on Computing in Education reports that "the pedagogical approaches ... for the development of valuable learning environments are still far from being implemented in most educational Web sites" (Mioduser et al. 2000). In this study, the researchers analyzed the pedagogical characteristics of 436 Web sites that they classified as "web-based learning environments." They found, for example, that only 28.2 percent of the sites include inquiry-based activities, only 32.6 percent focus on analysis and inference processes, barely 5 percent focus on problem solving and decision making, and just 2.8 percent of the sites support any form of collaborative learning. The Illuminations Web site has been designed to offer resources that embody all these effective learning processes, as well as other recommendations in *Principles and Standards.*

In fact, the Principles themselves—Equity, Curriculum, Teaching, Learning, Assessment, and Technology—have helped shape Illuminations. For example, much of the site is constructed around mathematics investigations that are intended to engage all students in learning important mathematics. Many of the investigations and lesson plans are threaded together to present local examples of developing a coherent curriculum of connected concepts and strands of mathematics. Resources such as classroom video vignettes and guiding questions for teaching and assessing are presented so that the investigations and lessons offer opportunities for teachers to reflect on the practice of teaching. Technology, of course, is integrated throughout the site and serves as a backbone to the project. These points will be elaborated and illustrated below, beginning with the following discussion of the different types of Illuminations resources.

Illuminations Online, Interactive, Multimedia Resources

The Illuminations Web site contains online, interactive tools for teaching, learning, and professional development in mathematics. The different types of online tools include (*a*) general purpose tools such as a spreadsheet, a grapher, and a shape manipulator; (*b*) content-specific tools, such as applets that facilitate the exploration of linear regression or fractions; (*c*) context-specific tools, such as applets that simulate light intensity or a game of pool; and (*d*) professional reflection tools, such as online video vignettes of students learning and teachers teaching mathematics. All these tools are integrated into mathematical or professional development activities that can be used to help achieve the vision of *Principles and Standards.* These online activities furnish examples of effective uses of the tools. The examples might be used directly in classrooms, workshops, or individually; they might be modified in many ways; or they might stimulate ideas for entirely new activities using the same or different tools.

For example, consider an interactive math applet that is tied to the specific context of a game of pool (illuminations.nctm.org/index_o.aspx?id=125). The pool game applet tool is embedded into an online mathematical investigation in a way that is in-line and just-in-time. That is, the applet is in place and ready to click and use when you need it. In the investigation, middle school students are asked to determine the ending corner, number of hits, and length of the path when a cue ball is hit at a 45° angle from one corner of a pool table.

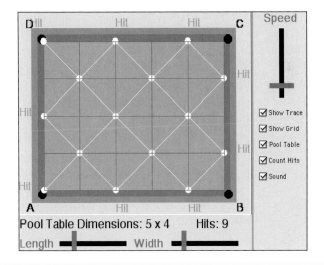

Pool Game Applet-Tool

This task supplies a rich context for students to further their understanding of ratio, proportion, greatest common factor, least common multiple, and symmetry. Although the investigation can be accomplished without technology, the applet significantly enhances teaching and learning by reducing the physical difficulties of carrying out the investigation, reducing the amount of time needed to gather necessary data, offering an easy method for testing conjectures and getting immediate feedback, increasing student ownership of the problem, and connecting a concrete problem to an abstracted grid representation.

Depending on the purpose and resources available, the applet-based pool-game activity can be used in the classroom in many ways. For example, this could be (1) a whole-class activity where a single computer is used to demonstrate and verify students' ideas, (2) a small-group activity in a computer lab, (3) an out-of-class activity in which students experiment and gather data that are discussed in subsequent classes, or (4) an end-of-unit project.

Other Illuminations tools are used to create online, interactive, multimedia professional development activities. These activities are currently built around video vignettes of teachers teaching and students learning mathematics. For example, the professional development activity Teaching, Learning, and Communicating about Fractions (illuminations.nctm.org/reflections /3–5/fractions/index.html) uses classroom videos of students playing a fraction game. The fraction game itself can be played online using two different applets in a one-player (illuminations.nctm.org/tools/fraction/fraction.asp) or two-player version (standards.nctm.org/document/eexamples/chap5/5.1 /index.htm). The classroom video vignettes show teachers modeling good communication skills as they encourage their students to communicate mathematically and think carefully about the fraction concepts. The videos and associated activities help teachers develop their ability to pose questions that elicit, extend, and challenge students' thinking, which is an essential part of creating a classroom environment in which intellectual risks, sense making, and deep understanding are expected.

As an example that uses online applets and videos together with off-line mathematical tools, consider Shedding Light on the Subject: Function Models of Light Decay (illuminations.nctm.org/index_o.aspx?id=137). This investigation includes an applet-tool simulating light intensity during an underwater dive, an off-line CBL experiment, an applet or use of calculators for analyzing data, video clips of students engaged in the investigation, teacher notes, and solutions. This activity also offers an example of how technology can influence what mathematics we teach, in this example more statistics and discrete mathematics, a functions approach to algebra, and more real-world mathematical modeling. See figure 15.2.

Illuminations Internet-Based Lesson Plans

Now we discuss the second major type of content found on Illuminations: Internet-based lesson plans. Illuminations lessons show how the Internet can be used for *Standards*-based mathematics lessons. For example, Internet links can furnish real-world data that can be analyzed or used to develop mathematical concepts. Other links may supply detailed information about areas in which mathematics is applied. Some links offer tools that can be used to graph, visualize, or compute. In all examples, the Internet links are used to enhance students' learning and promote more effective teaching.

A primary message that we endeavored to convey and build into the les-

Fig. 15.2. Overview of the Illuminations Shedding Light online investigation

sons is that "in planning individual lessons, teachers should strive to organize the mathematics so that fundamental ideas form an integrated whole" (NCTM 2000, p. 15). Toward that end, Illuminations lesson plans are in fact most often unit plans composed of five to seven sequenced lessons that develop substantial mathematical ideas across longer time periods. Because the plans have the complementary purposes of providing teaching resources, offering professional development for teachers, and communicating the vision of *Principles and Standards*, they are carefully crafted to present important mathematical ideas and effective instructional practices.

All lessons include questions to guide the development of mathematical understanding, suggestions for ongoing assessment experiences and proce-dures, a variety of instructional strategies, and questions to guide teachers' reflection. The unit plans demonstrate the value of technology as a tool for advancing and enhancing students' mathematical knowledge and their ability to use mathematics.

Notably, the unit plans are designed to accomplish the following:

• Illuminate and communicate the new vision of school mathematics presented in NCTM's *Principles and Standards for School Mathematics*

- Ensure access to *Principles and Standards* for many teachers at many levels of knowledge and experience

- Be examples of how to use the Internet in the classroom, taking advantage of the interconnected, interactive, multimedia nature of the Internet

- Offer opportunities for teachers at different levels of knowledge and experience to experience *Standards*-based teaching

- Furnish examples of what's important in a *Standards*-based lesson, including guiding questions, reflection activities, and sound and significant mathematics

- Facilitate effective classroom practice

For example, a unit plan for prekindergarten to grade 2 entitled "How Many More Fish?" (illuminations.nctm.org/index_o.aspx?id=51) consists of seven sequenced lessons that engage students in actively investigating five meanings and representations for the operation of subtraction: counting, sets, number line, balance, and the inverse of addition. See figure 15.3.

Illuminations Reviewed External Web Resources

The Internet is certainly valuable as a vast collection of information and resources. But this value quickly turns problematic as the quantity of information grows along with the time required to find what one is looking for. The Illuminations Web site particularly addresses this opportunity and problem through its collection of reviewed external Web resources (illuminations.nctm.org/swr/index.asp). In brief, this collection of resources gives *Standards*-focused access to the vast Internet virtual library.

These reviewed resources have four principal characteristics that make them useful for *Standards*-based mathematics education. First of all, the resources are organized according to the Standards and Expectations of *Principles and Standards*. Second, they are selected on the basis of a rigorous review by mathematics educators and mathematicians. Third, a written review of each resource is presented, which describes the resource and why it is useful for effective teaching and learning of mathematics and lists possible caveats that should be considered. And fourth, the resources are chosen at a small grain-size, often just one page on a larger Web site, focused on a particular topic and grade, so that they provide high-quality Web resources to help teachers teach specific topics at specific grade levels.

An objection that might be raised to this collection of resources is that it promotes a "bits and pieces" approach to the curriculum. Our view is that the resources we collect should be regarded as sources of information as the teacher weaves a coherent, focused curriculum. They should not be used as a curriculum in itself.

Internet-Based Multi-Day Lesson Plan / Pre-K - 2 ────────────

How Many More Fish?

Pre-K - 2   Number and Operations   Unit Plan   7 Lessons   30 minutes each

---

Overview: This unit plan focuses on comparative subtraction. The students use fish-shaped crackers to explore all five meanings for the operations of subtraction (counting, sets, number line, balance, and inverse of addition). In these lessons, the students use reasoning, then represent and communicate their findings.

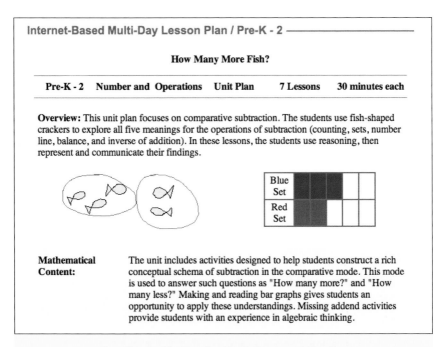

| **Mathematical Content:** | The unit includes activities designed to help students construct a rich conceptual schema of subtraction in the comparative mode. This mode is used to answer such questions as "How many more?" and "How many less?" Making and reading bar graphs gives students an opportunity to apply these understandings. Missing addend activities provide students with an experience in algebraic thinking. |

Fig. 15.3. An Illuminations unit plan on subtraction

Here is an example of how this collection could be used. Suppose a first-grade teacher will be starting on a unit on place value.

- *Step 1.* She selects "Number and Operations" for "Pre-K–2" from the navigation matrix in which the reviewed external resources are organized, as shown in figure 15.4.

- *Step 2.* She is given a list of descriptions of the available resources, which she then scans to see what might be appropriate. A partial list is shown in figure 15.5.

- *Step 3.* She clicks on the fourth item to learn more. She is given the following review (fig. 15.6) of the site, which suggests it might meet her needs, so she visits the site to see it for herself. The review also points out that Java is needed, so she checks with the resource center to be sure the computers in the lab have Java.

This example illustrates the most global search of the external Web resources collection. This procedure is the front end and navigation that users first encounter, so that they get a friendly, holistic vision of the collection. However, one might want to do a more targeted search. One could use a generic search engine, but instead we have developed what is essentially a *Standards*-based search engine. This system is an example of "contextualized

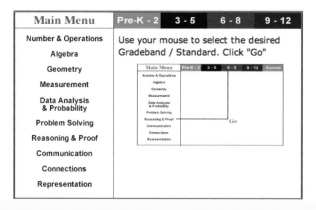

Fig. 15.4

Tour of Fractions (elementary level) - An extensive collection of lesson plans, FAQs, and software related to fractions for elementary school teachers. Direct to SWR

Orange Slices - Kindergarten students analyze the number of slices in an orange, introducing data analysis. Direct to SWR

Calculator Pattern Puzzles - Students use a calculator to explore some interesting patterns. Direct to SWR

Base Ten Blocks - This site gives a visual representation of base-10 place value and multi-digit operations, a potentially valuable tool for teachers to use in class or for students to use on their own. Direct to SWR

Fig. 15.5

navigation," which is discussed in more detail in the Web design section below. The advantage of this system is that it categorizes and searches the collection in an efficient manner that is directly tied to and helps illuminate *Principles and Standards.* Here is how it works for the same first-grade situation we considered above:

- *Step 1.* Start with the following search entry: *place value*

- *Step 2.* This step generates the following "key phrases" related to place value, which are taken directly from the Standards and Expectations of *Principles and Standards:*

Selected Web Resource: <u>Base Ten Blocks</u>	Standard: Number and Operation Grade Band: Grades PreK - 2 Resource type: Online Activities

Resource Review

This Web site consists of a Java applet, where one can select any combination of three different block sizes (representing a unit, 10 units, and 100 units) and drag them into the working panel. Students can then move, rotate, break, and glue the blocks to explore base 10 place value; older students in this grade band can explore multi-digit addition and subtraction. You can begin with the <u>instructions</u> or jump directly to the <u>applet</u>.

While virtual manipulatives may never replace the use of physical materials, there are some advantages. First, students can actually break apart the virtual blocks to decompose them into smaller blocks or glue collections of 10 blocks together to make larger blocks. With physical blocks, one has to "trade" a collection of blocks for another block (or v.v.) Second, the system constrains the gluing and breaking of blocks, so that incorrect regroupings cannot occur. The feedback may help students stay on track as they begin to explore place value and combining multi-digit numbers. This site could be appropriate either for individual student use (if appropriate technology, tasks, and guidance are available) or as a tool for teachers to lead a full-class discussion.

Fig. 15.6

- ✦ Base-ten blocks
- ✦ Comparing numeration systems
- ✦ Extending base-ten understandings
- ✦ Extending place value
- ✦ Introducing the base-ten numeration system
- ✦ Introducing place value
- ✦ Understanding place value

- *Step 3.* Choose the last of these, "Understanding place value." This relates to the Standards and Expectations as follows:
  - ✦ Standard: Number and Operations (Pre-K–2)
  - ✦ Goal: Understand numbers, ways of representing numbers, relationships among numbers, and number systems
  - ✦ Expectation: Use multiple models to develop initial understandings of place value and the base-ten number system

- *Step 4.* The "Understanding place value" key phrase generates the following list of Illuminations reviewed external Web resources that are assigned to that key phrase:
  - ✦ A Fictional History of Place Value
  - ✦ Base Ten Blocks
  - ✦ Counting the Rice

- ✦ Introduction to Place Value with Corn
- ✦ Place Value
- ✦ Scribble Square

- *Step 5.* Of these, the second, Base Ten Blocks, is the same resource used in the previous example, and the user gets the same review shown above.

Using Illuminations with Preservice Teachers

We will now present two brief data-based illustrations of how Illuminations has been used for professional development. In this and the next section we describe how Illuminations has been used with preservice and in-service teachers, respectively.

Principles and Standards recommends that 2-D and 3-D spatial visualization and reasoning are core skills that all students should develop. For example, students in grades 3–5 "should become experienced in using a variety of representations for three-dimensional shapes" (NCTM 2000, p. 169), such as isometric drawings, a set of views (e.g., top, front, right), and building plans. Spatial visualization has been defined as the "comprehension and performance of imagined movements of objects in two- and three-dimensional space" (Clements and Battista 1992, p. 444).

Isometric drawings have long been recognized as being difficult for both students and preservice teachers. Some of the sources of this difficulty include the inherent use of perspective, the projective nature of isometric drawings, which, for example, results in hidden lines and cubes and multiple realizations of any drawing, unfamiliar orientation, that is, the lack of "standard" x-, y-, and z-axes, and translation from isometric drawings to other representations, such as top-right-front views.

To help achieve the recommendation from *Principles and Standards* and overcome typical difficulties with isometric drawings, an isometric drawing applet (illuminations.nctm.org/tools/isometric/isometric.asp) was created in which students could explore the complexities of isometric drawings. The applet we created allows students to create dynamic drawings of three-dimensional objects on an isometric-dot grid (see fig. 15.7). Using this applet, the teacher can create isometric drawings with cubes, faces, or edges. Students can rotate, shift, color, decompose, and view them in 3D or 2D. In addition to the main isometric view, students can click the "eye" icon in the top menu bar to see two other views—the front-right-top views (also called "shadow views") that consist of three particular 2-D projections, and the mat plan or "building plan" view (see fig. 15.8).

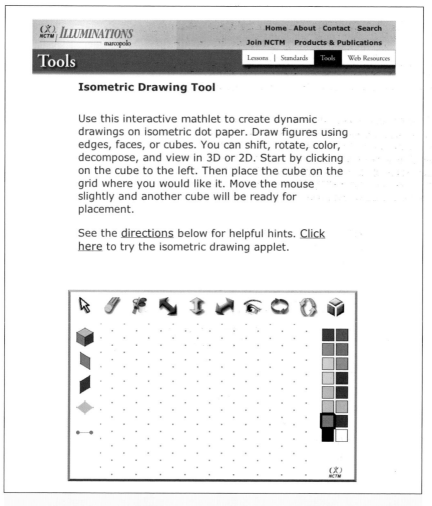

Fig. 15.7. The applet screen shot for the isometric drawing materials (figure continues on next page)

Also, objects built in the main window can be rotated in a new window (see fig. 15.9). This new rotated view is dynamically linked to the isometric drawing so that changes made in the object are automatically reflected in the new window. This feature is helpful when investigating problems associated with hidden cubes and other issues of projection.

For example, consider "impossible figures"—Escher-like drawings that can be interpreted as being impossible to build (see fig. 15.10). By rotating, students can get a new "illuminating" perspective on these figures.

Directions:

A HELPFUL HINT: It is best to draw your shape from the back to front and from bottom to top.

There are two ways of moving cubes. (1) You can SELECT and DRAG using the ARROW 🠦 option. The object will move only in the direction of the 3D axis. (2) Use the unit movement 🠦 ⬍ 🠦 buttons in one direction along the axes. You can click on the top or bottom of these buttons to move forward or backwards in the given direction.

Click on the VIEW 👁 button to see (a) Front-Right-Top views, (b) a mat plan, or (c) a 3D view which can be rotated using either the mouse or sliders.

Click on the ERASER ▱ button twice to erase all objects drawn.

Use the PAINT BRUSH 🖌 to color cubes, faces or edges. To color just one face of a cube, use the SHIFT key together with the paint brush.

Use the ROTATE BASE ⟳ or ROTATE VERTICALLY ⟲ to rotate your object.

Use the EXPLODE ⬢ button to change all cubes into faces. If two cubes share a face (i.e., the face is interior to the object), the face is not shown.

Fig. 15.7 (continued)

Several features of the applet have the potential to improve students' spatial skills and understanding of 2-D representations. Students can use the applet to create and reason about many more drawings and views than if the drawings are done by hand on paper. This is a common benefit of using technology and should facilitate the development of skills and understanding. However, we believed that the applet must accomplish more than just this potential benefit. Thus, in addition, the applet is designed so that objects can be drawn using cubes rather than using line segments as with paper draw-

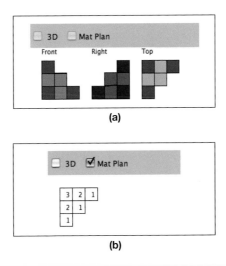

(a)

(b)

Fig. 15.8. Front-right-top views (a) and mat plan view (b) included in the applet

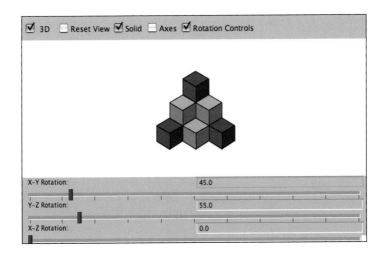

Fig. 15.9. Screen shot of a window for rotating a constructed object

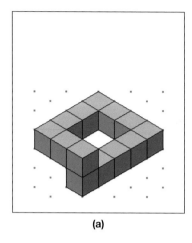

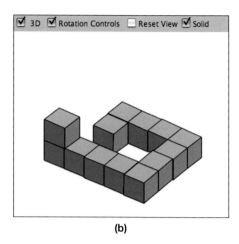

(a)   (b)

Fig. 15.10. An "Escher-like" figure, one in which the "natural" interpretation contains some contradiction; imagine walking along the top of the figure.

ings, thereby promoting a direct connection to the physical manipulation of concrete blocks. Also, the cubes can be moved without the limitations of physical space. Finally, multiple representations and views are furnished, including isometric views, front-right-top views, mat plans, and rotations through different axes. Moreover, these representations are all dynamically linked, so that whenever the user changes one view, the effect on the other views is immediately shown. The educational value of multiple linked representations has often been discussed and documented (e.g., in the Representation Standard in *Principles and Standards*). The use of technology, and this applet in particular, greatly facilitates creating and translating among multiple representations.

To help achieve the potential educational benefits of all these features of the applet, we created a set of online instructional tasks that engage students in using the applet to develop their spatial visualization skills. That is, we created an online unit by embedding the applet in rich curriculum materials. Rather than develop materials from scratch, and to maximize the usefulness of these materials, we have linked our Illuminations development work to prominent curriculum development projects with which we are collaborating. For example, the Connected Mathematics Project, which is a middle school curriculum development project funded by the National Science Foundation and designated Exemplary by the U.S. Department of Education, includes a geometry unit entitled "Ruins of Montarek" (Lappan et al. 2002). In this unit, students explore two-dimensional views of three-dimensional objects, including front-right-top views, isometric views, and mat plans. However, noticeably absent from these materials is the use of technology. We worked with the authors of this widely used middle school curriculum to develop an online applet-based enhancement of this unit.

We have used this curriculum-embedded applet as part of a course for pre-service teachers. Sixty undergraduates (two sections) worked through this unit as part of a geometry course for future elementary school teachers at a large state university in the Midwest. For example, students used the applet to help them draw an isometric representation. Toward this end, students could be seen holding their drawings to the screen to make comparisons. Of course some students could record their results without needing to refer back to the screen; they sometimes referred to this process as drawing "from memory" (see fig. 15.11).

Students were evaluated on a written test with five targeted categories: Correct Drawings, Shading, Build from Front-Right-Top views, Visualization, and Awareness. The responses on the written tests were evaluated as positive, negative, or not informative. A student was considered "proficient" in a category if 80 percent of the items related to that category were evaluated as positive. Figure 15.12 shows the percent of students reaching proficiency in the five targeted categories.

This discussion presents just a brief preliminary evaluation of the effectiveness of the applet-based approach to teaching and learning about isometric drawing. Our primary evaluative goal was to determine if learning was occurring in the intended categories. Overall, as figure 15.12 suggests, students' ability did improve in all five categories.

Using Illuminations with In-Service Teachers

In this section, we present an example of how Illuminations has been used with in-service teachers. Illuminations was used to furnish professional development for middle grades teachers participating in an Eisenhower project at a large southern state university. Relying on the vision set forth in *Principles and Standards for School Mathematics,* this project proposes to support teachers' professional development, including their knowledge of mathematics, their ability to assess learning, their leadership among colleagues, and their skill in integrating technology.

During the summer of 2001, three of the authors of this article facilitated a three-day workshop for thirty local middle school mathematics teachers. The workshop used Illuminations resources to provide professional development

Fig. 15.11. Two students' methods of recording the results of an exploration—one relying on the screen and the other recording "from memory"

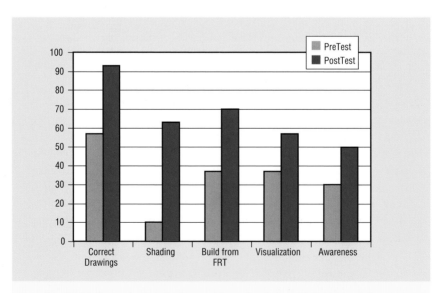

Fig. 15.12. Percent of students reaching a level of proficiency by category

related to the *Standards* documents for grades 6 through 8. Later in the summer, participants developed plans for implementing ideas from the workshop into their classes. During the academic year, these plans were carried out with support and monitoring from the project staff. The teachers and project staff recognized the potential of Illuminations applets and lesson plans for developing mathematical content knowledge, for improving the design and delivery of lessons, and for increasing students' interest and achievement in mathematics.

The dynamic, visual nature of the applets gave teachers a "picture" of mathematical concepts. For example, the slider on the fraction applet shows proportional relationships among fractions. This feature caused the teachers to rethink ways to help students (and themselves) understand proportionality and the part-whole relationship of fractions. Guiding questions, which are an important feature of Illuminations lessons, helped teachers consider the types of questions that would help focus students' attention on making sense of the mathematical concepts and the skills and applications associated with those concepts. When reviewing the format and sequence of an Illuminations lesson, the teachers recognized that a thoughtful presentation of the lesson in a logical sequence helped students recognize and make connections that build understanding. The idea of including assessment as a regular part of the instructional process was new to many teachers, yet by using the Illuminations lessons, they saw the value for students—for example, to help students build ownership of their own learning by explicitly perceiving their growth toward understanding the mathematical learning targets. The teachers also saw the value of monitoring the learning of their students throughout the unit, since this allowed them to change instructional activities more effectively, plan grouping strategies so that students could learn more effectively from one another, and know which students needed remediation at each step of the way. And finally, through the reflection activities that are part of many Illuminations resources, the teachers began to see the value of, and engage in, the process of regular reflection on their teaching in regard to its impact on their students' learning.

Conclusion

In this article, we have described the main types of Illuminations content, discussed the principles guiding the development of the content and the Web site design, and given examples of how Illuminations has been used for professional development. The Illuminations Web site continues to change and grow, but it maintains its focus on illuminating the recommendations for improving the teaching and learning of all students put forth in NCTM's *Principles and Standards for School Mathematics*.

REFERENCES

Clements, Douglas, and Michael Battista. "Geometry and Spatial Reasoning." In *Handbook of Research on Mathematics Teaching and Learning,* edited by Douglas A. Grouws, pp. 420–64. Reston, Va.: National Council of Teachers of Mathematics, 1992.

Lappan, Glenda, James Fey, William Fitzgerald, Susan Friel, and Elizabeth Phillips. "Ruins of Montarek: Spatial Visualization." A geometry unit in the Connected Mathematics Project. Glenview, Ill.: Prentice Hall, 2002.

Mioduser, David, Raft Nachmias, Orly Lahav, and Avigail Oren. "Web-Based Learning Environments: Current Pedagogical and Technological State." *Journal of Research on Computing in Education* 33, no. 1 (2000): 55–76.

Moor, Jane, and Rina Zazkis. "Learning Mathematics in a Virtual Classroom: Reflection on Experiment." *Journal for Computers in Mathematics and Science Teaching* 19, no. 2 (2000): 89–115.

National Council of Teachers of Mathematics (NCTM). *Principles and Standards for School Mathematics.* Reston, Va.: NCTM, 2000.

WebTrends Reporting Center. "Illuminations Web Site Monthly Report October 2003." MarcoPolo Partner Intranet, 2003.

16

Mathematics Teaching and Learning Supported by the Internet

Enrique Galindo

ARE you interested in taking advantage of technology to support mathematics learning but do not know what resources are available? Do you wish you could have access to interactive computer software to support students' construction of mathematical ideas? Are you interested in engaging your students in an integrated mathematics and science investigation as they collaborate with students in other schools? Do you have a pressing mathematics question that you would like to discuss with an expert? Do you want to learn more about current issues in mathematics education? Are you interested in networking with others who are working to improve mathematics learning? Do you wish your students could have access to real data to work on meaningful mathematics projects? Are you interested in learning more about high-quality mathematics teaching as you reflect on videotaped examples of teaching practices? These and many other exciting possibilities are now a reality, thanks to the Internet.

There are many ways in which the Internet is supporting and helping enhance the learning and teaching of mathematics today. In this article, some of the opportunities afforded by the Internet to both educators and students are examined using five broad categories, namely: resources, access to data, collaborative projects, video, and authoring. These categories are not meant to be exhaustive, and there may be some overlap among them, but they offer a framework for considering different ways in which mathematics educators can take advantage of Internet resources.

Resources

This first category consists of Internet sites that contain information, lesson plans, or curriculum materials to support mathematics learning and teaching. The first kind of resources in this section includes professional organizations and other resources to support the professional growth of those who

---

Editor's note: The CD accompanying this yearbook contains an electronic version of this article, complete with hyperlinked URLs.

want to improve mathematics education. The second group in this section includes resources to support mathematics learning through interactive applications. The last subsection furnishes entry points to lesson plans and other curriculum materials on learning mathematics topics for all grade levels and also on topics related to mathematics teacher education.

Professional Information

The Internet is an invaluable tool to those who are interested in current issues in mathematics education, want to learn about possibilities to network with others interested in the learning of mathematics, or want to have access to professional information related to mathematics learning. The Web sites in this section include professional organizations and other resources to support the professional growth of those who want to improve mathematics education.

The National Council of Teacher of Mathematics (www.nctm.org) is the world's largest professional organization of mathematics educators, with nearly 90,000 members. Throughout its activities NCTM offers vision and leadership to support mathematics teachers. This vision is articulated in *Principles and Standards for School Mathematics* (NCTM 2000). The messages set forth in this document come to life online through (1) its Web-based version (www.nctm.org/standards) and (2) the Illuminations project (illuminations.nctm.org). NCTM also provides leadership through its journals and other publications (www.nctm.org/publications), its professional development activities (www.nctm.org/academy), and its annual, regional, and other conferences (www.nctm.org/meetings).

The Math Forum (www.mathforum.org) is a great starting point for information about mathematics and mathematics education on the Internet. The Math Forum was started in January of 1996, thanks to a grant from the National Science Foundation, to investigate the viability of a virtual center for mathematics education on the Internet. Since then, the Math Forum has been a leading center to offer resources, materials, activities, person-to-person interactions, and educational products and services that enrich and support mathematics teaching and learning. Although the Math Forum includes resources that fall in many of the categories used in this article, the resources more closely related to professional information are briefly described here. Entry points to many important issues in mathematics education, such as constructivism, mathematics education reform, or the NCTM *Standards*, can be found among the Math Forum's collection of mathematics education resources (www.mathforum.org/mathed). Another area contains links about research on mathematics learning, including a section on summaries and discussions of seminal articles in mathematics education (www.mathforum.org/learning.math.html). Virtual interactions or collaboration with others are supported through public discussions (www.mathforum.org/discussions) and the Teacher2Teacher section (www.mathforum.org/t2t). Professional development opportunities include Math Forum workshops (www.mathforum.org/workshops) as well as workshops offered by others (www.mathforum.org/teachers/workshops).

Established in 1992, the Eisenhower National Clearinghouse for Mathematics and Science Education (www.enc.org/) supplies a selection of mathematics and science education resources on the Internet and supports teachers' professional development in mathematics, science, and the effective use of technology, among other activities. Two important features of the ENC Web site are the section devoted to support professional development (www.enc.org/professional) and the section on education topics, a section that helps locate articles that cover some of today's important topics for mathematics and science educators and parents (www.enc.org/topics).

The EQUALS programs at the Lawrence Hall of Science, University of California at Berkeley (www.lawrencehallofscience.org/equals), offer professional development opportunities and curriculum materials in mathematics and equity. A group devoted to support the reform of school mathematics presents information and networking opportunities through the Mathematically Sane Web site (www.mathematicallysane.com). Other professional organizations with an Internet presence include the Association of Mathematics Teacher Educators (www.amte.net), the American Mathematical Society (www.ams.org/), and the Mathematical Association of America (www.maa.org/). Web addresses for selected resources on professional information are listed in table 16.1.

Interactive Resources

One of the important ways in which the Internet can be used to support mathematics learning is through interactive applications. These resources use short programs written in the Java language called applets, or animations using Flash or other similar technologies, to engage learners interactively in mathematics investigations. Some of the first good examples of the possibilities for mathematics learning offered by these resources are the electronic examples, or e-examples, from Principles and Standards for School Mathematics (standards.nctm.org/document/eexamples). Consider, for example, the following investigation for the middle grades.

In the e-example titled "Understanding Congruence, Similarity, and Symmetry Using Transformations and Interactive Figures" (standards.nctm.org/

Table 16.1
Resources about Professional Information

Professional Information
National Council of Teachers of Mathematics www.nctm.org
The Math Forum www.mathforum.org
Eisenhower National Clearinghouse www.enc.org
Equals and Family Math www.lawrencehallofscience.org/equals
Mathematically Sane www.mathematicallysane.com
Association of Mathematics Teacher Educators www.amte.net
American Mathematical Society www.ams.org
Mathematical Association of America www.maa.org

document/eexamples/chap6/6.4), students use interactive figures to manipulate a shape and observe its behavior under a particular transformation or composition of transformations. In one part of this investigation, students are asked to determine the transformation that has been applied to a shape by comparing the shape to its image. The interactive figures allow students to choose different shapes and observe the behavior of their images, change the shape of a square or a triangle into other shapes by dragging from an edge or a vertex, or change the orientation of a shape by dragging from a vertex. Students can test their conjectures by generating alternative images using different transformations (see fig. 16.1). The goal is to provide students with opportunities and tools to deepen their understanding of transformations as they tackle the challenges presented in the investigation. The example includes information about the related NCTM Standards, suggests additional tasks, and highlights important issues for discussion and reflection.

In the e-example described above, thanks to the interactive figure, students can learn about the nature of unknown transformations by investigating the dynamic behavior of a shape and its image under the transformation and by paying attention to the invariants. That is, looking at the relationships that remain constant between the original shape and its image.

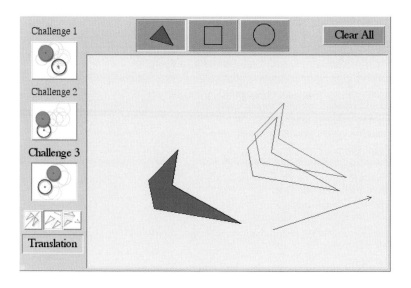

Fig. 16.1. Investigating an unknown transformation in the e-example titled "Understanding Congruence, Similarity, and Symmetry Using Transformations and Interactive Figures" (standards.nctm.org/document/eexamples/chap6/6.4)

The software supports students' learning as they gain new insights into transformations.

Additional examples of these interactive resources can be found in the Tools section of the Illuminations Web site (illuminations.nctm.org/tools/index.aspx). The Tools in the Illuminations site are built using Java applets and can be used to design investigations for students.

An excellent starting point when looking for interactive resources is the Math Tools project (www.mathforum.org/mathtools). Started in 2003, this project of the Math Forum, part of the National Science Digital Library (www.nsdl.org/), is a digital library of mathematical software essential to the learning of school mathematics. The library includes software for handheld devices, small interactive Web-based tools such as applets, and other computer software. In addition to reviewed technology tools, this library includes education research articles on relevant topics, offers support for teachers' use of math tools, supplies feedback for developers producing software meeting teachers' needs, and supports fruitful interaction for researchers of educational technology. Additional entry points to interactive resources on the Internet are listed in table 16.2.

Table 16.2

A Selection of Interactive Resources for Mathematics Learning and Teaching

Interactive Resources
Principles and Standards E-examples standards.nctm.org/document/eexamples
NCTM Illuminations: Tools illuminations.nctm.org/tools/index.aspx
Math Tools www.mathforum.org/mathtools
National Library of Virtual Manipulatives www.matti.usu.edu/nlvm
Project Interactivate www.shodor.org/interactivate
Educational Software Components of Tomorrow (ESCOT) project www.escot.org
Arcytech www.arcytech.org/java
Count On www.mathsyear2000.org
Mathematical Sciences Digital Library www.mathdl.org
Mathemagica Math Tools www.jason.org/eprise/main/jason_public/web_pages/mathemagica/math_tools/ math_tools.htm
CLIME Microworlds www.mathforum.com/clime/microworlds
Manipula Math with Java www.ies.co.jp/math/java
Alive Maths hydra.educ.queensu.ca/cgi-bin/Maths/maths.cgi
Descartes 3 descartes.cnice.mecd.es/ingles
NCES Students' Classroom nces.ed.gov/nceskids
University of Alberta GAMES Group cs.ualberta.ca/~games

Curriculum Materials

A wealth of curriculum materials can be found on the Internet; they range from short activities and lesson plans to complete Web-based courses on different topics (see table 16.3). Materials can be found on the Internet on topics on the learning of mathematics for all grade levels, as well as on topics related to mathematics teacher education. A good entry point to find mathematics lessons and other curriculum materials on the Internet is the Math Forum's Teachers' Place (www.mathforum.org/teachers). Resources on this

Table 16.3
Selected Entry Points to Curriculum Materials for
Learning and Teaching Mathematics

Curriculum Materials
The Math Forum Teacher's Place www.mathforum.org/teachers
NCTM Illuminations Project : Lesson Plans illuminations.nctm.org/index.aspx
NCTM Illuminations Project : Selected Web Resources illuminations.nctm.org/swr
Annenberg/CPB Teacher Resources: Mathematics www.learner.org
PBS TeacherSource www.pbs.org/teachersource/math.htm
PBS TeacherLine teacherline.pbs.org/teacherline
InterMath intermath-uga.gatech.edu/homepg.htm
EdWeb www.edwebproject.org

Web page are organized by grade level. Several special interest topics such as adult education or science fair math projects are included. This Web page also offers a self-guided tour to features and services of interest to educators.

Two areas from the Illuminations site (illuminations.nctm.org/) are of special relevance when looking for curriculum materials. The section about lesson plans (illuminations.nctm.org/index.aspx) contains a collection of Standards-based lesson plans that take advantage of resources on the Internet. The Illuminations site also compiles a collection of selected Web resources (http://illuminations.nctm.org/swr). The Web resources in this section have been reviewed and approved by the site's editorial board because they met three criteria: the mathematics is accurate, the resource fosters understanding or supports the implementation of Principles and Standards for School Mathematics (NCTM 2000), and the resource is well organized, easy to navigate, and well maintained. Another useful source of curriculum materials on the Internet is the PBS Teacher Source (www.pbs.org/teachersource/math.htm).

In addition to resources for students, educators can find on the Internet curriculum materials to support their professional growth. The multimedia resources found at the Annenberg/CPB Web site (www.learner.org/) provide excellent opportunities for professional development. Teachers can access educational video programs with coordinated Web and print materials. The video programs can be viewed online, accessed by schools from a satellite channel, or ordered on videotape. In addition to multimedia resources on teaching mathematics, five college-level mathematics courses for elementary and middle school teachers can be found in the Learning Math area of this Web site (www.learner.org/learningmath). Other entry points to professional development resources are the PBS Teacher Line (teacherline.pbs.org) and the InterMath site (www.intermath-uga.gatech.edu/homepg.htm). Finally, those interested in the role of the Web in education will find many resources on the EdWeb site (www.edwebproject.org/).

Access to Data

Every time we need to make a decision today, whether about business, politics, or everyday life, we can find statistical information to help us make that decision. It has become more important than ever before that we have the knowledge and skills to interpret and analyze such information. The data available on the Internet can help educators develop learning experiences to support students as they learn to reason statistically. Furthermore, the data available on the Internet can help support mathematics learning in areas other than statistics while helping students make connections between mathematics and other school subjects, and between mathematics and everyday situations. Two types of Web resources related to data are examined in this section: real-time data projects and projects using existing data sets. Table 16.4 lists a selection of Web sites related to data.

Real-Time Data Projects

In this type of projects, real-time data from the Internet are used to engage students in an investigation. Projects are usually multidisciplinary, so it is not uncommon to address subjects such as marine science, chemistry, physics, biology, mathematics, and language arts in the same investigation. These projects may also use other sources of data in addition to the real-time data. Examples of real-time data available on the Internet include an air-quality index from the U.S. Environmental Protection Agency (www.epa.gov/airnow), earthquake activity from the U.S. Geological Survey (www.earthquake.usgs.gov/), weather information from the Interactive Weather Information Network (iwin.nws.noaa.gov/iwin/main.html), satellite images from the National Oceanic Atmospheric Administration (www.noaa.gov/satellites.html), and current marine data from Oceanweather Inc (www.oceanweather.com/data).

Table 16.4
Selected Data–Based Internet Resources for Learning and Teaching Mathematics

Access to Data

Real-Time Data

CIESE Real Time Data Projects
k12science.stevens-tech.edu/realtimeproj.html

Musical Plates Project
k12science.stevens-tech.edu/curriculum/musicalplates3/en

Air Quality Index from EPA
epa.gov/airnow

Earthquake Activity from USGS
earthquake.usgs.gov

Weather Information
iwin.nws.noaa.gov

Satellite Images
www.noaa.gov/satellites.html

Marine Data
oceanweather.com/data

Data Sets

The Data Library
mathforum.org/workshops/sum96/data.collections/datalibrary

InfoNation
cyberschoolbus.un.org/infonation/info.asp

Quantitative Environmental Learning Project
seattlecentral.org/qelp

Exploring Data
exploringdata.cqu.edu.au

StatLib
lib.stat.cmu.edu

Statistical Reference Datasets
www.itl.nist.gov/div898/strd

The Center for Improved Engineering and Science Education (CIESE) has developed a number of classroom projects that take advantage of real-time data. Teachers in grades K–12 throughout the world can use these interdisciplinary projects to enhance their curriculum, taking advantage of the Internet. The projects are also a good source of ideas and resources to create new projects. In addition to student activities, the CIESE Real Time Data Projects (k12science.ati.stevens-tech.edu/realtimeproj.html) include links to reference materials, a teacher's guide, and other support materials.

An example of a real-time data project is the Musical Plates project (k12science.ati.stevens-tech.edu/curriculum/musicalplates3/en) for students in grades 6–12. In this project students use real–time earthquake and volcano data from the Internet to explore the relationship among earthquakes, plate tectonics, and volcanoes. Retrieving information from the U.S. Geological Survey, students can find the location, depth, and magnitude of earthquakes for a recent period, say, the previous two weeks. Information from NASA's Jet Propulsion Laboratory helps estimate the motion of Earth's tectonic plates for a given time period, say, a million years. Combining this information with data on volcanic activity, students look for patterns and relationships among these three events. They can formulate a hypothesis and develop a plan to test their hypothesis. Students can learn to represent information graphically, interpret information furnished in a graph, look for patterns and trends and draw conclusions from collected data, and use mathematics to sum up information and convey it to others. Figure 16.2 shows the home page for the Musical Plates project.

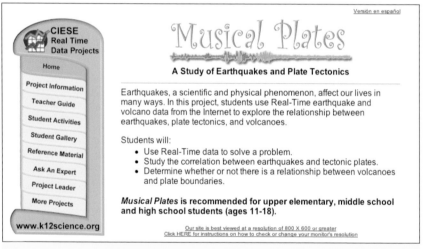

Fig.16.2. Home page for the Musical Plates project (k12science.ati.stevens-tech.edu/curriculum/musicalplates3/en)

Data Sets

Many interesting investigations can also be designed around existing data. The Internet gives access to a number of collections of data sets that can be used in mathematics class. A good starting point is the Math Forum's Data Library (www.mathforum.org/workshops/sum96/data.collections/datalibrary). This data library contains several data sets that can be downloaded and used in a spreadsheet. The data library also supplies links to other Web sites with data sets.

A brief description of the contents of other data-related Web sites follows. The InfoNation Web site (www.cyberschoolbus.un.org/infonation/info.asp) contains statistical data for all United Nations member states. Users of this site can easily select two or more countries to display comparisons for several statistical data. The Quantitative Environmental Learning Project (www.seattlecentral.org/qelp) furnishes resources to integrate mathematics and environmental science in college classrooms. Teachers of introductory statistics can find curriculum support materials in the Exploring Data Web site (exploringdata.cqu.edu.au/). StatLib (lib.stat.cmu.edu/) is a Web site for distributing statistical software, data sets, and information. Finally, the Statistical Reference Datasets Web site (www.itl.nist.gov/div898/strd) offers reference datasets with certified computational results to use for the objective evaluation of statistical software.

Collaborative Projects

Communication in the mathematics classroom is a powerful means to help students develop mathematical understanding. When students are asked to communicate about the mathematics they are studying, explain their reasoning, justify their solution methods, or describe aspects of the problematic situation they do not understand, they gain insights into their mathematical thinking. When working on meaningful problematic investigations with others, students' opportunities to communicate increase and have more relevance. The Internet can facilitate meaningful collaboration among individuals or groups to support mathematics learning. Groups or individuals can collaborate with experts at a distance; entire mathematics classes can participate in joint explorations with students in other cities, or even other parts of the world. Collaboration among students in different geographic locations or different countries adds a richer dimension to the collaboration. These collaborations can provide access to information or resources that otherwise would be unavailable; they can also help foster an understanding of our lives and our world from a mathematical perspective. The ways in which the Internet can support two types of opportunities for collaboration are described in this section: group projects, and consulting with experts. Table 16.5 has a selection of Web sites for collaborative projects.

Group Projects

An example of a collaborative project supported by the Internet is the Connecting Mathematics to Our Lives Project (www.orillas.org/math). In this project, students in classrooms around the world are invited to collaborate in

Table 16.5

Selected Internet Resources for Collaborative Projects for Learning and Teaching Mathematics

Collaborative Projects

Group Projects

Connecting Mathematics to Our Lives
www.orillas.org/math

Journey North
www.learner.org/jnorth

Noon Day Project
k12science.org/noonday

Project Groundhog
www.stemnet.nf.ca/Groundhog

Class2Class
mathforum.org/class2class

Global Schoolnet's Global Schoolhouse Organization
globalschoolnet.org/GSH

BROWSE Lybrary of Inquiry Projects
cleo.terc.edu/cleo/browse/template/
browse-homepage.cfm

CIESE: Collaborative Projects
k12science.stevens-tech.edu/collabprojs.html

Consulting with Experts

Ask Dr. Math
mathforum.org/dr.math

GoMath.Com On-Line Math Help
gomath.com

AskNRICH Ask a Mathematician Service
nrich.maths.org/discus/messages/board-topics.html

Quandaries and Queries
mathcentral.uregina.ca/QQ

Scientific American Ask the Experts
sciam.com/askexpert_directory.cfm

activities that relate curriculum content to students' individual and collective experience and to use mathematics to analyze educational and social issues relevant to students' lives. Figure 16.3 shows the home page for this project.

Teachers and students participating in the Connecting Mathematics to Our Lives Project are offered a range of activities. Activities such as What Mathematics Means to Me, Statistics in Society, and Promoting Equity at Our School Site challenge students to think critically and to act on what they are

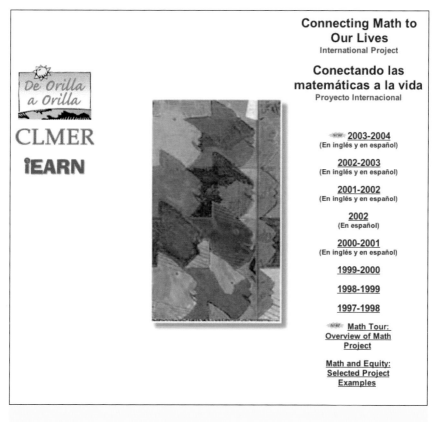

Fig. 16.3. Home page for the Connecting Mathematics to Our Lives Project (www.orillas.org/math), a collaborative global project

learning through democratic participation in their schools and communities. In one of the activities related to promoting equity at the school site, second and third graders from a class in the United States partnered with students in Spain to analyze biographies in their school's and city's libraries. Students were to categorize biographies on the basis of gender, race, class, or disability. Students learned about percentages, fractions, bar graphs, and other mathematical means to make sense of data as they described the libraries' biography collections. They also had to wrestle with complex social issues as they worked with some of the categories, for example, race, which students found difficult to distinguish from color. Second-language learning was also part of this experience, as some students became familiar with some of the terms used in the other language.

Another example of a collaborative project supported by the Internet is Journey North (www.learner.org/jnorth), a project that engages students in a global study of wildlife migration and seasonal change. Students in grades K–12 share their field observations about migration patterns of wildlife and changes in natural events, such as the budding of plants and sunlight, with classmates across North America. Participants in this project learn important mathematics and science concepts as they make predictions about migration, explore the concept of time and its relation to biological clocks, explore maps and location and their relation to the sense of location of some animals, and use nature's clues to predict the arrival of spring. Other examples of collaborative projects are the Noon Day Project (k12science.ati.stevens-tech.edu/noonday), where students recreate Eratosthenes' measurement of the earth's circumference, and the Project Groundhog (www.stemnet.nf.ca/ Groundhog), which engages young learners in exploring whether or not the groundhog is an accurate weather predictor.

Several Web sites are good entry points to collections of collaborative projects. The Math Forum's Class2Class site (www.mathforum.org/class2class) was created to facilitate student and class participation in Internet projects. The Global Schoolhouse (www.globalschoolnet.org/GSH) has a clearinghouse of more than 900 online collaborative projects. BROWSE (cleo.terc.edu/cleo/browse/template/browse-homepage.cfm) is a library of mathematics and science inquiry projects maintained by TERC. CIESE has developed a good collection of collaborative projects (k12science.stevens-tech.edu/collabprojs.html).

Consulting with Experts

An interesting possibility offered by the Internet is that of submitting mathematics questions to experts. Several Web sites allow teachers or students to submit questions. The Math Forum's Ask Dr. Math (www.math-forum.org/dr.math) is a question-and-answer service for mathematics students and their teachers. It includes a searchable archive and a list of frequently asked questions. The GoMath Web site (www.gomath.com/) offers free online mathematics help for grades K–12 within twenty-four hours. Cambridge University in England offers the Ask A Mathematician service (nrich.maths.org/discus/messages/board-topics.html), where a team of university students answers mathematical questions. The University of Regina in Canada offers the Quandaries and Queries Web site (mathcentral.uregina.ca/QQ), a question-and-answer service provided by their Department of Mathematics and Statistics and their Faculty of Education. Science as well as mathematics questions can also be submitted to the Ask the Experts section of the Scientific American magazine (www.sciam.com/askexpert_directory.cfm).

Video

As more teachers gain access to high-speed Internet connections, their access to video on demand is becoming a possibility. Having access to video-taped lessons or video clips of episodes of mathematics learning offers excellent opportunities for professional growth. Video of vignettes of mathematics learning, properly supported, can be an excellent tool for teachers to reflect on effective teaching—a tool that, thanks to the Internet, can be available to teachers anytime, anywhere. Professional development opportunities based on video can be supported with links to background information, lesson plans, students' work, commentary from teachers and experts, teachers' reflections, and moderated discussion groups.

One example of this type of online professional development is the Seeing Math project (seeingmath.concord.org/). Building on the case method of learning, which is widely used in schools of business, law, and medicine, this project has added the support of audio, video, and interactive Web materials to develop powerful Web-based video case studies. Each of the Seeing Math video case studies focuses on specific mathematics content that is deemed difficult to teach and presents a particular pedagogical strategy to provide online professional development for elementary and middle school teachers. Cases are supported with video commentaries of (1) the teacher reflecting on the lesson taught and (2) mathematics content specialists and specialists from the field bringing different perspectives to help participants use the experience in the video example to understand other learning situations. Support materials help answer questions about the school, students, lesson plan, students' work, assessment, and curriculum. Another interesting component of this project is the related development of digital tools. For example, the project has developed a tool called VideoPaper Builder, which enables users to create their own video cases.

Another example of online professional development based on video is the Inquiry Learning Forum (ILF) (ilf.crlt.indiana.edu/). The ILF seeks to improve students' learning in mathematics and science by supporting teachers in better understanding inquiry-based teaching and learning. The ILF uses a school metaphor. When logging in, users are presented with several "rooms" typical of a school building. Within these virtual rooms, one can obtain or share lesson plans, view videotaped classes of fellow teachers, participate in online discussions, and collaborate online with groups focused on a particular topic. The videotaped classes are linked to a number of support materials, such as lesson plans, students' work, teachers' reflections, and issues for discussion. The goal is to present a vehicle for discussing teaching practice. Through observation, discussion, and reflection, participants can find their own paths to professional growth and development.

The Annenberg/CPB mathematics programming, a collection of video libraries and instructional video series on different topics of grades K–12 mathematics learning and teaching, is another opportunity for professional development that takes advantage of online video. Educators can access videos and other Web-based materials from the Annenberg Web site (www.learner.org/). Another interesting path for professional development is the Lesson Study method developed in Japan. Lesson study is a process in which teachers jointly plan, observe, analyze, and refine actual classroom lessons called "research lessons." Videotaping research lessons is useful to help analyze them at a later time. Educators interested in learning about this method can find video, articles, and other resources at the Web site of the Lesson Study Group at Mills College (lessonresearch.net/). Table 16.6 has a selection of Internet resources that use video for professional development.

Table 16.6
Selected Entry Points to Professional Development
Supported by Video Available on the Internet

Professional Development Internet Resources Based on Video
Seeing Math seeingmath.concord.org
Inquiry Learning Project ilf.crlt.indiana.edu
Annenberg/CPB Projects learner.org
Lesson Study Group at Mills College lessonresearch.net

Authoring

The Internet offers teachers and students the opportunity to publish their work and make it accessible to the world. Authoring is a high-level cognitive activity, and students can learn a great deal from it. Furthermore, when publishing is combined with opportunities to receive feedback about the published work, authors benefit from the suggestions and ideas they may receive from others reading or using the published work. Some of the options available to teachers and students to publish their work are described in this section. Table 16.7 lists a selection of resources related to this category.

Table 16.7
*Selected Internet Resources for Publishing Work
Authored by Teachers or Students*

Authoring Internet Resources

ThinkQuest
thinkquest.org

Teacher Exchange
mathforum.org/te

ESCOT: Interactive Problems
escot.org/resources/problems/overview.html

Create Your Own Interactive Games
oswego.org/staff/cchamber/techno/games.htm

The Oracle Foundation funds a semiannual competition to encourage students between the ages of nine and nineteen to develop creative Web sites. Students work in teams of three to six students, supervised by a teacher-coach. Students have about five months to develop their Web site, focused on a given topic area. Sites that meet the evaluation criteria are published in the ThinkQuest Web site (thinkquest.org/). The site has created a library with the lessons resulting from these competitions; it is a free public collection of lessons created by students for students.

Teachers interested in publishing their lessons can use the Math Forum's Teacher Exchange area (www.mathforum.org/te). The Teacher Exchange allows teachers to submit links to lessons they have created, or to create lessons that will be hosted at the Math Forum. Lessons are then added to the collection, which is organized by grade bands and content standards. Teachers interested in creating simple interactive matching games can use a resource made available by the Oswego City School District in New York (oswego.org/staff/cchamber/techno/games.htm).

The Educational Software Components of Tomorrow (ESCOT) project was funded to research how teams can compose lessons by combining graphs, tables, simulations, algebra systems, notebooks, and other tools available from a shared library of reusable components. Teams worked thorough online collaboration to author challenging interactive problems (escot.org/resources/problems/overview.html).

Using Internet Resources to Support Students' Learning

Educators interested in using the Internet to support students' learning will find many examples of resources and ideas in the previous sections of this paper. However, how do we incorporate these resources into the mathematics classroom? One model to design learning experiences based on Internet resources is the WebQuest. A WebQuest is an inquiry-oriented activity in which some or all of the information that learners interact with comes from resources on the Internet (Dodge 1995). WebQuests offer opportunities for students to engage in investigations that are based on Internet resources and that promote students' learning as they are challenged to use higher-order thinking. Many guidelines have been developed to help educators design WebQuests. Two good starting points to learn more about WebQuests are The WebQuest Page (webquest.sdsu.edu/) and The WebQuest Portal (webquest.org/).

The basic WebQuest typically contains six sections: Introduction, Task, Process, Resources, Evaluation, and Conclusion. The Introduction section seeks to motivate students, pique their interest, and explain what the investigation is about. The final product that is expected is described in the Task section. The Process section describes steps students will take to accomplish the task, and it sometimes assigns roles to students. The Resources section includes the Web pages and other resources that are needed to complete the project. The Evaluation section outlines how the final product and the process are going to be assessed. The Conclusion section is an opportunity for students to review and apply what they learned.

Five guiding principles to have in mind when creating WebQuests have been outlined by Dodge (2001). One of them is "Challenge learners to think." This principle is to remind us that it is very easy to focus on searching for resources or for information when using the Internet. However, what matters is what the students do with those resources or that information. Another principle is "Use the medium." This principle encourages us to consider what makes the Internet different from other media. If it is access to multimedia resources or the ability to interact with others, then we should seek to take advantage of those unique opportunities in the Web-based investigations we design for students.

Selecting Internet Resources

It is useful to have some criteria in mind when making decisions about selecting and using Internet resources to support mathematics learning and teaching. A brief list of criteria that might be useful to consider follows. Goldenberg (2000) outlines six principles to use when deciding what are good uses of technology in mathematics learning. Because his principles are about new technologies in general, they are very relevant when making decisions about using Internet resources in mathematics class. Paul Goldenberg's six principles can be rephrased as follows.

1. *Consider the level of thinking that the learning situation requires when choosing a genre of Web resource.* An important finding from research on the use of technology in mathematics education is that what makes a difference in students' learning is how technology is used rather than the use of technology per se. The history of using technology in mathematics education shows that when using a new technology tool to support mathematics learning—for example, interactivity in a Web resource—there is a strong temptation to use that new tool to support low-level thinking skills, say, the drill and practice of basic facts. However, technology has been more effective to support learning of higher-order thinking skills, such as reasoning and problem solving. We may see more impact on student learning if we choose Web resources on the basis of whether they support the learning of high-level thinking skills.

2. *Consider the purpose of the lesson.* A learning situation such as a lesson or extended investigation should have a clear purpose, and all learning resources used should help students focus their attention on that main goal. For example, let us say middle grades students are working on a lesson whose purpose is to help students learn about linear functions by having students examine relationships among distance, time, and speed. In such lessons, having students work with paper and pencil, doing the calculations by hand, or trying to examine many instances of the situation may not be feasible and actually may be an obstacle to students' understanding of the relationships involved. However, having students work with a dynamic simulation of two runners along a track, where students can set the speeds and starting points of the runners, run the simulated race, and examine graphs of time versus distance to look for relationships that describe this situation mathematically, can be a very appropriate use of a Web resource for the purpose of that lesson.

3. *Consider the nature of the thinking required.* There are times when what is important is to pay attention to the steps of the process, to the detail, and to all factors involved, and we do not want the Web resource to obscure that detail. If students are learning about finding relationships in data, for exam-

ple, using a Web resource that furnishes access to real-time data may give students opportunities to work with a more meaningful real situation, consider the specific aspects of the data, and face the challenge of modeling an authentic situation, whereas using a Web-based simulation may not help students realize what are all the variables that need to be considered when looking for relationships in that context. The Web resource that is selected should not obscure, or bypass, the thinking that is required in a given learning situation.

4. *Consider the thinking skill we want to develop.* The tools we use to support students' learning should not replace or hinder the thinking skills that we consider important for students to develop. Rather, tools or resources should foster the development of important thinking skills. If we choose to use interactive figures to help students learn about transformations of geometric shapes, for example, it is not because we seek that students will now need the interactive figures every time they have to predict the outcome of a given transformation. We need to use the Web resource in ways in which it can help students gain a deep understanding of transformations—the type of understanding that will enable them to perform mental experiments with transformations, and predict the outcomes of a composition of transformations, even in the absence of the interactive figures.

5. *Consider the content of the curriculum carefully.* Internet resources offer many interesting opportunities to rethink the mathematics curriculum. Students may now have access to the study of mathematics topics we could not have considered in the classroom before. Easier opportunities for integration with other subject areas are brought within teachers' reach. Collaboration opportunities we could not have envisioned before are now possible. However, with so many possibilities there are also risks. We can easily send students on a Web-based scavenger hunt where the mathematics content gets lost. We need to remain critical when we consider making changes to the curriculum in light of the availability of Web-based resources. We need to keep the learning of important mathematics at the center of our goals.

6. *Consider a developmental approach to the incorporation of Internet resources.* As with any tool to support mathematics learning, it takes time for teachers and students to learn to use Internet resources effectively. In addition to the learning considerations described above, there are many practical aspects that need to be considered. Practical factors, such as the type of setting available—computer lab or a few computers in the classroom, the ease of access, and the hardware and connection speed available—can present additional challenges with which teachers and students need to learn to deal. It may take a long time for teachers and students to become powerful users of the Internet to support mathematics learning. It is important to bear in mind

the time and opportunities that are needed for educators and students to become savvy users of these new resources.

References

Dodge, Bernie. "Some Thoughts about WebQuests." Retrieved from edweb.sdsu.edu/courses/edtec596/about_webquests.html, 1995.

———. "FOCUS: Five Rules for Writing a Great WebQuest." Retrieved from babylon.k12.ny.us/usconstitution/focus-5 rules.pdf. 2001.

Goldenberg, E. Paul. "Thinking (and Talking) about Technology in Math Classrooms." Retrieved from www2.edc.org/mcc/iss_tech.pdf. 2000.

National Council of Teachers of Mathematics (NCTM). *Principles and Standards for School Mathematics.* Reston, Va.: NCTM, 2000.

17

Laptops in the Middle School Mathematics Classroom

Suzanne Lewis

WHAT if your school implemented a laptop computer program, and as a teacher, you had to integrate laptop technology into your mathematics classroom? Many mathematics teachers face exactly this challenge. Their schools have adopted laptop technology, but they don't always have a defined strategy for integrating laptops into the curriculum. Instead, the teacher must design a curriculum that uses laptops effectively.

I teach eighth-grade algebra in a pre-K through grade 12 school that has been using laptop technology for eight years. Each teacher has a laptop, and all students in grades 5–12 have their own laptops. It took me two years to learn how to use the computer and the software programs effectively in the classroom, even though I had past computer experience. Even now, I long for more time to explore ways in which I can use laptop technology to enhance my students' learning. The purpose of this article is to help readers understand why computers, and especially laptops, enhance the mathematics classroom and to offer suggestions and examples of how to integrate laptops into their mathematics curriculum.

By using spreadsheets and other software on their laptops, students can easily generate data, which allows us to shift the focus from the actual calculations to seeing patterns and developing mathematical models. An example of this is work I do with decimals and percents in an exercise called "A Living Wage." The students use a spreadsheet to convert an hourly wage into pay per day, per week, per month, and per year; they then write formulas that take out taxes. With the results, they can do repeated subtractions on the spreadsheet as they calculate what it costs to live after paying for rent, food, transporta-

---

Editor's note: The CD accompanying this yearbook contains an electronic version of this article, complete with hyperlinked URLs. It also contains two lessons in Word format and associated Excel files that are relevant to this article.

tion, clothes, utility bills, insurance, and so on. The students can change the amount of the wage in one cell, and the remainder of the spreadsheet changes accordingly. This allows the students to focus on the questions "What is a living wage?" and "How can a person live on minimum wage in our country?" (See fig. 17.1.)

Imagine what it would be like to have a job that paid minimum wage or close to it.
Minimum wage is $5.15 an hour. It was last raised by Congress in September 1997.

	Hour	Day	Week	Month	Year
Minimum Wage	**$5.15**	$41.20	$206.00	$824.00	**$9,888.00**
Actual Take-Home Wage	**$4.05**	$32.39	$161.97	$647.87	**$7,774.44**
Minus Soc.Sec. 7.5%					
Minus 15% for taxes					
Monthly Wage	**$824.00**				
Monthly Wage after taxes	**$647.87**	$647.87			
House/Apt. Payment	$300.00	$347.87			
Car Payment	$0.00	$347.87	no car		
Gas and Car Insurance	$0.00	$347.87	no car insurance		
Food	$300.00	$47.87			
Heat, Electricity	$80.00	-$32.13			
Clothing, shoes	$50.00	-$82.13			
Medical	$50.00	-$132.13			
Vacations, Entertainment	$20.00	-$152.13			
Total Expenses	**$800.00**				
Difference	**-$152.13**				

Fig. 17.1

The students see how difficult it is to live on minimum wage. Using a spreadsheet also teaches the students how they can make a budget, buy a car, calculate house payments, and keep track of investments in the stock market.

Since the entire spreadsheet changes after just one cell is changed, students can answer questions such as "If the minimum wage was changed to $7.00 an hour, could you afford a car and car insurance?" (See fig. 17.2.) With an extra $80.60, you can't afford a car yet. How much would you need to earn before you can buy a car? The students can keep replacing the minimum wage number and determine what hourly wage they would need to earn to be able to buy a car. (See fig. 17.3.) With laptops, the students are able to take them home to work on these activities and share their findings with their families.

Laptops offer students immediate access to the Web without leaving the classroom to go to a computer lab. They can easily access interactive Web sites

Imagine what it would be like to have a job that paid minimum wage or close to it.
Minimum wage is $5.15 an hour. It was last raised by Congress in September 1997.

	Hour	Day	Week	Month	Year
Minimum Wage	**$7.00**	$56.00	$280.00	$1,120.00	**$13,440.00**
Take Home	**$5.50**	$44.03	$220.15	$880.60	**$10,567.20**
Minus Soc.Sec. 7.5%					
Minus 15% of what's left for taxes					
Monthly Wage	**$1,120.00**				
Monthly Wage after taxes	**$880.60**	$880.60			
House/Apt. Payment	$300.00	$580.60			
Car Payment	$0.00	$580.60	No car		
Gas and Car Insurance	$0.00	$580.60	No car insurance		
Food	$300.00	$280.60			
Heat, Electricity	$80.00	$200.60			
Clothing, shoes	$50.00	$150.60			
Medical	$50.00	$100.60			
Vacations, Entertainment	$20.00	$80.60			
Total Expenses	**$800.00**				
Difference	**$80.60**				

Fig. 17.2

where they can investigate equations, build geometric figures, see how fractals are made, and make connections to the real world. In my classroom, students use many interactive Web sites. For example, there are several sites where students see the graph of a line change as they enter different numbers for the slope and y-intercept in an equation. In a few minutes the students can visually interpret negative slope, positive slope, steepness, and the y-intercept. Pencil-and-paper graphing is done so the students know how to graph ordered pairs and basic equations using slope and intercept, but the students then can move on to the simulations and the analysis of the slope/intercept form of the equations. Several years ago, prior to our laptop program and the ability to go to interactive Web sites in the classroom, students would laboriously graph by hand and analyze graphs. With the laptops and the Internet, the students can see what happens immediately as the numbers change and can analyze the equations. With wireless projectors, the students can project their work onto a screen for everyone to see. The appropriate use of the laptop—in conjunction with student activities, classroom culture, and the teacher's role—enhances the student's mathematical learning. Why laptops instead of desktops? We found that when each student owns his or her own laptop, it levels the playing field for all students. Many of my weaker math students have strong computer skills: they

Imagine what it would be like to have a job that paid minimum wage or close to it.
Minimum wage is $5.15 an hour. It was last raised by Congress in September 1997.

	Hour	Day	Week	Month	Year
Minimum Wage	**$10.00**	$80.00	$400.00	$1,600.00	**$19,200.00**
Take Home	**$7.86**	$62.90	$314.50	$1,258.00	**$15,096.00**
Minus Soc.Sec. 7.5%					
Minus 15% of what's left for taxes					
Monthly Wage	**$1,600.00**				
Monthly Wage after taxes	**$1,258.00**	$1,258.00			
House/Apt. Payment	$300.00	$958.00			
Car Payment	$0.00	$958.00	no car		
Gas and Car Insurance	$0.00	$958.00	no car insurance		
Food	$300.00	$658.00			
Heat, Electricity	$80.00	$578.00			
Clothing, shoes	$50.00	$528.00			
Medical	$50.00	$478.00			
Vacations, Entertainment	$20.00	$458.00			
Total Expenses	**$800.00**				
Difference	**$458.00**				

Fig. 17.3

shine in that regard. Prior to laptops, the students could use desktops at school in our lab, but it was difficult to schedule lab use when it was needed. Often the students would end up sharing a computer. Those students who had their own computers at home and had Internet access always had an advantage.

Another advantage to laptop use is that students who cannot make it to class can see the lessons they missed by looking at the class notes that I took on my laptop during the lesson and posted on my Web site or e-mailed from another student's laptop. Our newest laptops are tablets, where the screen swivels and the top can be used as a writing surface. Students can take notes, writing the math problems by hand and drawing pictures that illustrate their work (figures such as circles or solids) just as they do on paper, but they can save these notes as a document and then e-mail the notes to anyone. Teachers are doing lessons on tablets, saving their lessons and placing them on their Web sites, so students who are absent can see the lessons. Teachers have students take quizzes on their tablets and e-mail the quizzes directly to the teacher. The teachers correct them on their computers and e-mail them back to the students. Teachers can write class lessons by hand on the computers, project the notes on a screen or whiteboard instead of using the chalkboard, save the written notes, and publish them on their Web site for anyone to use.

Laptops allow students to use many different types of software in the classroom. Some ways that I use the standard applications in my classroom are as follows:

1. *Word Processing*

   - Students keep a math journal using word processing.
   - Students write reports on statistical analyses they complete using a spreadsheet.
   - Students learn to import tables and graphs from a spreadsheet into their word-processing documents.

2. *Spreadsheet*

   Students do more than twenty lessons using a spreadsheet. Examples are presented below.

3. *Presentation Software*

   Students prepare slide presentations related to mathematics and show them to the class. Students have selected topics including famous mathematicians, scientists, optical illusions, and the golden rectangle.

4. *Web-Design Software*

   Students are required to have a Web site and learn how to design their own. Students' Web sites must have a Web page for each class, so they have a portfolio of their work online.

5. *E-Mail*

   Students make online survey forms on their Web site for an eighth-grade math project. Students, faculty, and other adults take the surveys and submit the results directly through their e-mail—no more surveys on paper!

6. *The Internet*

   Students use the Internet to do research for projects. They also use lessons on the Internet, use interactive sites, and research the stock market for playing the Stock Market Game. In addition, students can use my Web page as a resource to find homework assignments, extra-credit problems, math class activities, and Internet links to a wide range of subjects.

7. *Scientific Notebook*

   I use Scientific Notebook as a tool to write mathematics lessons. I can then project the lessons onto a screen from my computer instead of using the chalkboard. I can also review the previous day's lesson with the students without having to write it out again, which saves valuable classroom time. At the end of the year, some of my students use Scientific Notebook to graph equations and explore some algebra 2 topics.

In the past, students would have completed many of these activities and assignments with pencil and paper. Other activities require the power of the computer to do what couldn't be done before. Following are some examples of student activities that use a spreadsheet or the Internet to complete assignments that couldn't reasonably be done before.

Using a Spreadsheet in the Mathematics Classroom

The program the students use most in eighth-grade mathematics is a spreadsheet that includes many mathematical functions, including statistics, graphing, and operations with numbers using formulas that are entered by the user. Students learn how to put different formulas into the cells to generate the results they want. After learning the basics of the spreadsheet, the students can work independently to find new ways to use it to solve problems. I have created more than twenty lessons that require the students to use a spreadsheet. A sample lesson is called Pythagorean Triples.

When we study rational and irrational roots, the students study the Pythagorean theorem as a way to understand the concept of square roots. There are many proofs of the Pythagorean theorem that are interactive on the Internet. Students learn best by visualizing and interacting, and the Internet allows them to do this. One activity I do to practice using the Pythagorean theorem and to help the students understand that in most right triangles the hypotenuse is irrational is to try to find Pythagorean triples (i.e., the set of positive integers satisfying the equation $x^2 + y^2 = z^2$). I ask the students to find as many Pythagorean triples as possible. The assignment allows them to use the Pythagorean theorem many times and also presents a challenge. Seven years ago, prior to laptops, students were able to find three triples at most by hand without the use of a spreadsheet. Patterns were difficult to find, and the students generated their answers with random guesses. Most students found 3, 4, 5 and the related family 6, 8, 10. Some found 5, 12, 13. Using a spreadsheet we make together in class, students can now generate many Pythagorean triples quickly. Figure 17.4 shows the spreadsheet the students design to find Pythagorean triples.

With the Pythagorean triples spreadsheet, the computer generates in less than a second results that take hours to generate with a calculator. Learning how to use a spreadsheet and seeing the speed at which the spreadsheet works are only part of the positive and creative experiences for the students. The stu-

Pythagorean Triples
To find other triples, just enter a different number into A7
and it will automatically change the whole table.

Formula in A7	Formula in B8	Formula in C7	|	Pythagorean Triples Families				
.=A$7	.=B7+1	.=sqrt(A7*A7+B7*B7)	3-4-5	5-12-13	7-24-25	8-15-17	9-40-41
A	B	C	6-8-10	10-24-26			
3	1	3.16227766	9-12-15				
3	2	3.605551275					
3	3	4.242640687					
3	4	5					
3	5	5.830951895					
3	6	6.708203932					
3	7	7.615773106					
3	8	8.544003745					
3	9	9.486832981					
3	10	10.44030651					
3	11	11.40175425					
3	12	12.36931688					
3	13	13.34166406					
3	14	14.31782106					
3	15	15.29705854					

Fig. 17.4. Spreadsheet to find Pythagorean triples

dents learn by seeing patterns and can then make predictions about what numbers can be Pythagorean triples. They can see all the possibilities and the relationships and understand the underlying mathematical principles more easily when they can see the results of their work in spreadsheet form.

Another example of using a spreadsheet is graphing and comparing equations. Students learn mathematical modeling from doing the problems, and the computer enhances their understanding of the power of exponents, since they can graph all the equations using the spreadsheet. Students can do this with a calculator as well, but with the laptop, they can present their findings in either a slide-show format or a word document with explanations. The students can see the data points next to each other and compare them more easily than on a calculator. The students are assigned four problems to solve by making an equation for each.

1. *Cost of a notebook:* You are going to the store to buy some notebooks. Each notebook costs $5.00. How much would you spend for 1 to 15 notebooks?

2. *Area of a garden:* You are asked to find the number of square feet in a square garden. Given that the possible lengths of the sides are 1 to 15 feet, determine the areas of the garden.

3. *Volume of a cube:* You are asked to find the volume of a cube. Given that the sides range from 1 inch to 15 inches, determine the volumes of the cube.

4. *Rabbit population:* Every three months, the rabbit population doubles in size. If you start with 2 rabbits, after 15 three-month periods, how many rabbits are there?

The students develop the four equations: linear, quadratic, cubic, and exponential. They then create the data points from the equations using a spreadsheet (see fig. 17.5) and make a graph to compare all four equations (fig. 17.6).

Other examples of lessons using a spreadsheet involve graphing, percents, statistics, decimals, exponents, square roots, the stock market, trigonometry, and more.

	Linear, Quadratic, Cubic, and Exponential Equations			
x	**Cost of Notebook** $y=x$ **Cost-Linear**	**Area of a Square** $y=x^2$ **Area-Quadratic**	**Volume of a Cube** $y=x^3$ **Volume-Cubic**	**Rabbit Population** $y=2^x$ **Rabbits-Exponential**
1	5	1	1	2
2	10	4	8	4
3	15	9	27	8
4	20	16	64	16
5	25	25	125	32
6	30	36	216	64
7	35	49	343	128
8	40	64	512	256
9	45	81	729	512
10	50	100	1000	1024
11	55	121	1331	2048
12	60	144	1728	4096
13	65	169	2197	8192
14	70	196	2744	16384
15	75	225	3375	32768

Fig. 17.5. Using a spreadsheet to create data points

Using the Internet in the Mathematics Classroom

The Internet offers many opportunities to explore mathematics. Because we emphasize looking for patterns in mathematics and problem-solving activ-

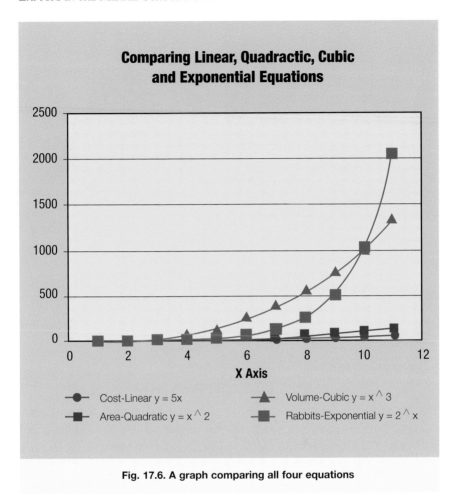

Fig. 17.6. A graph comparing all four equations

ities, I have a lesson on fractals with several pictures of fractals hanging in my classroom, including the Mandelbrot set. We discuss the pictures, looking at patterns. We go on the Internet and do a lesson on fractals looking at Sierpinski's triangle.

I give the students some Internet sites, and they find other sites on their own. The lesson asks them to draw their own Sierpinski triangle on paper. After students draw their own triangle, they then go to interactive sites where the computer draws a Sierpinski triangle. They can see the fractal develop before their eyes. One question that I asked is what fraction of the drawing is shaded, and the students make a chart to see the pattern of each subsequent drawing (see fig. 17.7).

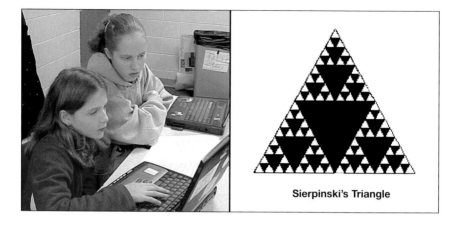

Sierpinski's Triangle

We often use the following links to resources to study fractals:

- Sierpinski Triangle Lesson—Cynthia Lanius
  (math.rice.edu/~lanius/fractals/inpr.html)
- Interactive Sierpinski Triangle—Arcytech
  (www.arcytech.org/java/fractals/ sierpinski.shtml)
- Interactive Sierpinski Triangle—Serendip
  (serendip.brynmawr.edu/playground/sierpinski.html)

Prior to the class, I put links to these sites on my Web page. When the students go there to see the math lesson for the day, they can follow the links and do much of the investigation on their own. I project the different sites on a screen, and we have conversations in small groups or with the whole class as the students read, draw, and do the interactive investigations.

Another example of using the Internet for mathematics education is playing the Stock Market Game. My students have played the Stock Market Game

Iteration	1	2	3	4
Number of shaded triangles	1			
Total number of triangles	4			
Fraction of triangles shaded	1/4			

Fig. 17.7

online for two years. In 2003, we took a record first, second, and third places in the state of Ohio. The Stock Market Game has only a few rules: Given $100,000 and a minimum of five stocks to purchase, each team of three students researches stocks online and trades stocks during a ten-week period. Each team has to decide which stocks to buy and how many shares to buy. Students buy and sell stocks with a 2 percent transaction fee for each buy or sell. Because they have their own laptops, they are able to do most of the work on their own time, once they are shown how to research stocks. The first-place team made a 28 percent profit in a ten-week period, winning a prize of $125. The Stock Market Game is a real-world experience. The students could do research on the Internet with ease and speed using the laptop computer. The students made a spreadsheet to record their stock transactions. They made graphs and kept track of their profits and losses. Doing research and making their own decisions allowed them to feel connected to the real world. Our goal for our students to be independent thinkers and researchers was enhanced because of technology and opportunities like playing the Stock Market Game.

Other examples of lessons using the Internet include the following ones, which are available on the CD accompanying this yearbook:

- Slope/intercept form of linear equations and their graphs: interactive site that allows you to change the slope and intercept and see the line shift
  - ✦ id.mind (id.mind.net/~zona/mmts/functionInstitute/linearFunctions /lsif.html)
- Quadratic equations and their graphs: interactive sites that allow you to change the values of the parameters of the equation to see how the graph of the parabola changes
  - ✦ PERSO (perso.wanadoo.fr/pilat/english/svg/parabole.htm)
  - ✦ IES.CO (www.ies.co.jp/math/java/conics/draw_parabola /draw_parabola.html)
- Graphing linear inequalities: interactive site that allows you to change the terms and the sense of the inequality
  - ✦ BigChalk (www.bigchalk.com/cgibin/WebObjects/WOPortal.woa/wa /HWCDA/file?fileid=298291&flt=ga)
- Stock Market Game and related sites:
  - ✦ Stock Market Game (www.smgww.org/)
  - ✦ Yahoo Finance (finance.yahoo.com/)
  - ✦ Hoovers (www.hoovers.com/)
  - ✦ Morningstar (www.morningstar.com/)

- Many interactive sites list:

  - ExploreMath
    (www.exploremath.com/activities/Activity_page.cfm?ActivityID=38)

Finally, in 2004–05 we are implementing online math textbooks for algebra 1 and prealgebra. There are several benefits of using an online textbook. The most significant benefit is that students can interact with many of the lessons. These interactions occur through applets that allow the student to change numbers and move items. For example, in one exercise the student can manipulate the coefficients of a quadratic equation and watch the corresponding changes in the graph of a parabola. Since the material is built-in, I don't need to search for applets on other Internet sites. The online textbook also includes video lessons, in which a teacher explains the skill taught in the lesson. Vocabulary definitions can be heard (in both English and Spanish) through an audio component using Real Player. There is also a significant cost advantage. A six-year online version for the book costs less than $4 per student per year. Finally, middle school students do not have to carry home a heavy textbook; they can just open the text online. Since we have online courses in history and grammar as well as algebra, we have eliminated three heavy textbooks. We do keep a classroom set of textbooks in case we can't connect online.

Final Comments

Students raised in the "digital age" are accustomed to seeing images change in one second. Don Tapscott (1999) calls today's students the "Net Generation" in his book, *Growing Up Digital & the Rise of the Net Generation.* Tapscott says on his Web site, www.growingupdigital.com, " 'N-Geners' are using digital media for learning. 60 percent of American households with children have computers and use them for learning a variety of things. These children, between the ages of two and 22, are hungry to learn new technologies. The implications are massive as family members begin to respect each other for their knowledge and the kids have the knowledge." For the first time ever, the students will be the teachers and the teachers will be the students.

In the mathematics classroom, the use of computers mirrors our society because most mathematics will be done with the assistance of computers. This generation of students will shop, bank, and invest online. It makes sense to expose them to the power of the computer in mathematics by using the software available to them to increase their understanding of, and facility with, mathematics. Spreadsheets and the Internet are two excellent ways to integrate technology with mathematics education. This connection with technology can be used to make mathematics in the classroom come alive.

Developing a strategic program that incorporates technology into the mathematics classroom is challenging and rewarding. I am always being stretched by computer technology, and I am always looking at how my students are thinking differently. I began using laptops designing one lesson at a time— and I still do. I try out new ideas and research how I can use technology to increase my students' understanding of mathematics. You will find ever-increasing ways to expand its use once you start using laptop technology in the classroom, see the benefits, and realize that your students are learning to love mathematics using computers.

REFERENCE

Tapscott, Don. Growing Up Digital: The Rise of the Net Generation. New York: McGraw-Hill, 1998. (www.growingupdigital.com/menu.html)

18

The Role of Technology in Representing Mathematical Problem Situations and Concepts

Dominic D. Peressini
Eric J. Knuth

CURRENT educational reform efforts in the United States are setting ambitious goals for schools, teachers, and students (e.g., National Research Council 1993; National Education Goals Panel 1991). Teachers are to help all students in grades K–12 learn to value content, become confident in their ability to solve problems in specific content areas, and learn to reason and communicate from content-specific disciplinary perspectives (National Council of Teachers of Mathematics [NCTM] 1989, 1991, 2000; National Council for the Social Studies 2000; National Council of Teachers of English 1996). Spanning across all these goals is the recommendation that meaningful uses of technology be incorporated in all areas (International Society for Technology in Education 2000). The new visions of technology-rich classrooms called for by national and state standards pose great challenges for teachers, since they represent a substantial departure from the K–12 classrooms in which most of today's teachers and prospective teachers were students.

The reformed visions of mathematics classrooms called for by the NCTM *Standards* documents (NCTM 1989, 1991, 1995, 2000), in particular, pose similar challenges for mathematics teachers and students, since they are based on fundamentally different assumptions about mathematics, about teaching and learning, about students and schools, and perhaps most important, about the role of technology in mathematics education from those on which much of teachers' existing professional knowledge is based. To move successfully toward these visions, teachers will need to develop, practice, and implement new forms of instruction and assessment that are grounded in the meaningful use of technology.

Editor's note: The CD accompanying this yearbook contains a Geometer's Sketchpad file that is relevant to this chapter. For more information on The Geometer's Sketchpad, contact Key Curriculum Press at www.keypress.com/.

Developments in communication and technological tools and structures (e.g., optical and digital media, the Internet and the World Wide Web, Internet television, powerful software packages, decreases in production costs of computer hardware) have literally changed industry, commerce, and society. Accordingly, schools must also change so that they prepare students to function successfully in this new information society. In addition, many of these technological devices (e.g., graphing calculators, computers) and software packages (dynamic geometry, dynamic algebra, ecological system simulators) can be used by mathematics teachers in their individual classrooms in order not only to familiarize students with technology applications in the domain of mathematics but also to facilitate students' learning of basic skills and algorithmic procedures, data collection and analysis, and conceptual knowledge in mathematics. Consequently, opportunities to introduce, learn, and experiment with a variety of technology must be provided to practicing as well as prospective teachers so that they can begin to integrate this technology into their mathematics and classrooms in meaningful ways. The purpose of this article is to explore the role of technology in the mathematics classroom and, in particular, to highlight the ways that individual teachers can use technology to foster students' understanding through the representation of mathematical tasks and concepts.

Technology in the Mathematics Classroom

During the past several years, we have been working with both in-service and preservice mathematics teachers in an effort to help them better understand, and implement, reform-based mathematics instruction and assessment. One of our central goals has been to assist teachers in their efforts to integrate technology into their own classrooms. From our experiences working with these teachers—and these teachers' students—we have formulated five primary ways in which technology is currently being used as a pedagogical tool in mathematics classrooms.

The first role technology plays in the mathematics classroom is as a *management tool* that helps teachers and students work more efficiently. For teachers, this includes mathematics-focused Internet home pages on the Web that contain ideas, tasks, activities, and actual lesson plans that teachers can download to use with their students. Other resources include software packages such as grade-book programs, spreadsheets, word processors, and databases that allow teachers to more efficiently organize, prepare, and modify multiple aspects of their classroom responsibilities and daily activities. In addition, hardware such as multimedia projectors, LCD panels, and other such technological tools helps teachers better structure and facilitate their classroom activities. Technology plays a similar management role for students in allowing them to fulfill some of their responsibilities as students more efficiently. For example, mathematics-focused Internet home pages on the Web can provide students access to a variety of information to aid them in explorations of the history of, or current research in, mathematics. In addition, students can access other sites that furnish mathematical tools like random-number generators, raw data for analysis, such as census data, and opportunities to interact with, and ask questions of, "mathematics experts." And like teachers, students may also use software packages such as word processors and spreadsheets to create, organize, and modify their work not only in mathematics but in other classes as well. In this capacity, we have also included the use of calculators and other such software that allows students to perform mathematics algorithms and procedures more efficiently and accurately.

A second role technology plays—as a *communication tool*—is to enable mathematics teachers to connect with other mathematics teachers both within and across schools so that they can collaborate and share their teaching successes and failures as well as discuss other teaching-related issues. This can be accomplished by both e-mail talk groups and Internet discussion forums; these dynamic and interactive media for posting teaching issues and the responses and reactions to the issues engender reflection and support continued discussion around them. Recent technological advances also make Internet observation of classroom video clips possible discussion topics. Further, both mathematics teacher educators as well as preservice mathematics teachers can participate in such virtual discussions. In addition, these technology-based communication tools can be used to foster discourse and collaboration among educators, students, parents, and the community.

In the third role, technology can be used as an *evaluation tool* that assists teachers in reflecting on their instruction and providing feedback for student learning. In this role, teachers may again use grading software and video observations to furnish feedback to individual students. Video observations may also be used for teachers to reflect on their own instruction and enhance what they are doing in their mathematics classrooms. In addition, e-mail, discussion boards, and interactive Web sites present more alternatives for teachers to discuss their teaching methods and get feedback from other mathematics teachers and educators around the world. Finally, recent advances in technological tool design allow classroom technology such as graphing calculators to be interconnected in such a way that a teacher can have direct access to each individual students' calculator screen. This can also be done in a computer-lab setting as students work on their own computers. Teachers can mirror each students' screen on their own computer in order to monitor students' work and progress. Access of this sort thus gives teachers opportunities to evaluate individual student's thinking, adapt instruction if necessary, or pose additional questions for students to explore.

Technology, in the fourth role, can be used as a *motivational tool* to encourage and engage students in learning mathematics. As we have all observed firsthand, technology in the form of graphing calculators, software packages, and other technological tools often engages students for the sole reason of the technology itself—that is, these types of technology are often novel in classrooms, and the use of a new and exciting pedagogy almost always engages students because of its uniqueness and, in the example of technology, because of its dynamic nature. In addition, technology also supplies a multiplicity of pathways for students to engage in the mathematics tasks at hand, and this variety of points of engagement allows more students to gain access to the mathematical procedures, ideas, and concepts. Furthermore, the interactive nature of many mathematics-based technologies allows students to become active participants in the mathematics classroom and may foster a sense of autonomy that both students and teachers often find refreshing. Finally, student access to technology allows teachers (as well as students) to pose interesting and motivating mathematics problems—problems that would be difficult to explore or solve without the use of technology (cf. Knuth 2002; Knuth and Peterson 2003).

The fifth role, and perhaps the most important from the perspective of school mathematics reform, is to harness technology in ways that help students better understand mathematical algorithms, procedures, concepts, and problem-solving situations. In this capacity, as a *cognitive tool,* technology offers a unique means for supporting students' exploration of, and engagement with, mathematics. It affords "new ways of representing complex concepts, and make[s] available new means by which students [or teachers] can manipulate abstract entities in a 'hands-on' way" (Perkins et al. 1995, p. xvi). Indeed, as we reflect on the way that we have used technology as a cognitive tool with the mathematics teachers with whom we have worked, we find ourselves engaged in two distinct but related processes. The first process focuses on using technology to help students represent mathematical tasks or problem-solving situations so that they more fully understand the context in which a particular task or problem resides and the important factors that influence the activity at hand. In this sense, technology allows teachers and students to "manipulate abstract entities in a 'hands-on' way" (Perkins et al. 1995, p. xvi). The second process—often set in the context of the first process just described—allows students to use technology to represent and explore a variety of mathematics procedures and concepts so that they can be examined from a conceptual perspective. In this fashion, technology serves the function of "representing complex concepts," so that students and teachers can more fully understand the conceptual underpinnings of a variety of mathematics content (Perkins et al. 1995, p. xvi).

In the remainder of this article, we examine two mathematics tasks (focusing on several central, important mathematical ideas) and some of the ways in which cognitive-technological tools can be used in the classroom to foster robust understanding of not only the problem situation itself but also the important mathematical ideas contained within the context of the mathematics task. In addition, we will return to the first four roles as we discuss some of the considerations regarding the use of technology in the mathematics classroom.

Representing Mathematical Problems and Concepts with Technology

This section gives an overview of two problems and how technology might be used as a cognitive tool for representing the problem situations and for exploring the mathematical ideas underlying each problem. The first problem, a TIMSS (Trends in International Mathematics and Science Study, formerly known as the Third International Mathematics and Science Study) problem, engages students in making conjectures about the areas of triangles that share a common base. The second problem, a calculus problem, involves a more abstract problem situation and deals with the complex idea of rate of change. Although the use of technology in exploring each problem certainly provides students with a means for dynamically representing the respective problem situations, its use in representing the latter problem, as will be seen shortly, plays a more pivotal role in furthering students' understandings of significant mathematical concepts. Accordingly, we shall argue that technological representations of this latter type need to be emphasized in mathematics instruction if technology is to be viewed as a powerful tool for thinking mathematically.

Representing the Relationship among the Areas of Triangles

This section describes a TIMSS geometry warm-up problem. The original problem did not permit the use of interactive geometry software as a means for arriving at a solution; however, the problem can easily be adapted for use with such applications. The use of technology to represent this problem then allows an empirically based exploration of the problem situation.

In the diagram in figure 18.1, segments *AB* and *CD* are parallel. Given this information, what conjectures can you make about triangles *HGF*, *EGF*, and *IGF*? In particular, what can you say about the relationship among the areas of these triangles. Can you generalize, and prove, your conjecture?

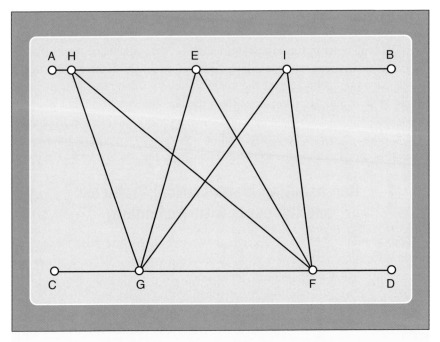

Fig. 18.1. A TIMSS geometry warm-up problem

Exploring this problem with interactive geometry software requires a construction that allows students to freely move both the locations of the three vertices, H, E, and I, along segment AB and the endpoints of the base, G and F, along segment CD. Such a dynamic construction enables students to examine the areas of a number of different triangles. With the use of the interactive geometry program's area measure function, it is likely that most students will arrive at the correct conclusion: the areas of the resulting triangles are equal. Although at this point students will not have proved that the areas are equal, the technological representation of the situation enables students to gain conviction regarding the truth of the conclusion. An important next step is that students actually engage in the process of proving to verify that the conclusion is indeed true; however, the reason for engaging in this next step goes beyond simply to verify the conclusion. The proof also furnishes valuable insight into the mathematics underlying the conclusion—the areas of the triangles are equal because the triangles all share a common base and their heights are also equal.

Representing a More Complex Problem Situation

This section will describe a series of questions used to enrich and extend a traditional calculus problem. The use of technology in exploring the problem affords students an opportunity to visualize the varying problem extensions in ways that may lead to a deeper conceptual understanding of the underlying mathematics. Further, in offering a means for visualizing the problem as well as its extensions, the use of technology also makes the underlying mathematical concepts more accessible to a wider range of students.

The following problem (Thomas and Finney 1979) is one typically found in many calculus texts: *A ladder 26 feet long is leaning against a vertical wall. The foot of the ladder is being moved away from the wall at a rate of 4 feet per second. How fast is the top of the ladder sliding down the wall at the instant when the foot of the ladder is 10 feet from the wall?* This problem can be modified in ways that make extensive use of technology. The technological representations used in exploring and in visualizing the "new" problems supply students with a tool for thinking about the different mathematical ideas that would be difficult at best without the use of technology. In what follows, we first present three questions (Monk 1992, pp. 181–82), each of which focuses on different, but related, aspects regarding the movement of the ladder. Next, the role technology might play in exploring the different problem situations is discussed, with a particular emphasis placed on the role technology might play in students' thinking about the problem and the associated mathematical ideas (see corresponding Geometer's Sketchpad files[1]).

Question 1. Tom sees the ladder in a position like the one shown in figure 18.2, almost vertical against the wall. He moves the bottom a small distance further away from the wall and records the amount by which the top of the ladder has dropped. Some time later Sally comes along and sees the ladder in a position like the one shown in figure 18.3. Sally moves the ladder away from the wall *by the same amount that Tom did* and records the amount by which the top of the ladder has dropped. Does Tom or Sally see the top of the ladder drop more? Or do they see the ladder drop equal amounts? Explain.

Question 2. Now, Katrina comes along and sees the ladder against the wall. She pulls the ladder away from the wall by a certain amount and then again by the same amount, and then again by the same amount, and then again, by the same amount, and so forth. Each time she does this, the top of the ladder drops down the wall a bit. As she does this, she records the distances by which the top of the ladder drops down each of these times. Do the amounts by which the top of the ladder drops remain constant as Katrina repeats this step; or do they get bigger, or do they get smaller? Explain.

---

1. We used Geometer's Sketchpad software in constructing the files that accompany this article; however, similar constructions could be produced using Cabri or using handheld devices such as the TI-89, TI-92 Plus, and TI Voyage 200, which are loaded with an interactive geometry program.

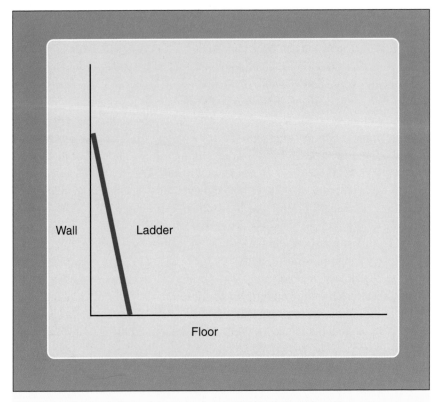

Fig. 18.2. The ladder in its initial position before Tom moves it

Question 3. Newt, the science nerd, then comes along and puts wheels on the bottom of the ladder. He connects them to a motor so that the bottom rolls away at a constant, but very slow, speed. The top of the ladder slides down at a very slow speed. Does the ladder move downward at a constant speed; or does it speed up, or does it slow down? Explain.

Prior to any attempts to determine answers to the preceding questions with the use of technology, students should be asked for their initial thoughts. The correct answers to the questions are somewhat counterintuitive; thus, the results may serve as a "hook" for engaging the students with the subsequent investigations. It would not be surprising for students to suggest initially that equal outward movements of the base will result in the top of the ladder dropping by equal amounts (in response to Question 1), that continued constant outward movements of the base will result in the top of the ladder dropping by constant amounts (in response to Question 2), and that moving the base of the ladder out at a very slow constant speed will result in the top of the lad-

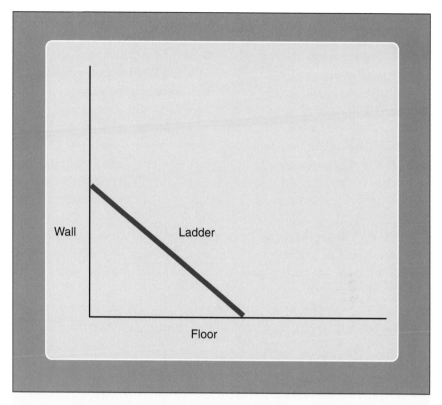

Fig. 18.3. The ladder in its initial position before Sally moves it

der moving downward at a constant speed (in response to Question 3). To actually determine an answer to the first two questions, students, using a dynamic construction, can move the bottom of the ladder outward by equal amounts and then record the distance that the top of the ladder drops after each movement, thus allowing direct comparisons of the amounts by which the top of the ladder dropped. In answering the third question, one can use Sketchpad's animation function to actually observe the effect of moving the base outward at a constant speed—one can "see" the top of the ladder dropping at an increasing rate of speed.

Up to this point, the students have either relied on Sketchpad-supplied measurements or on their own observation in determining answers to the three questions. The question of why the ladder behaves as it does, however, remains to be answered. To investigate this problem further—and to begin to shed light on the mechanism underlying the situations—students should trace the path that a point near the top of the ladder follows as the base is pulled away at a constant rate of speed (fig. 18.4). On the basis of this trace, it is possible to discuss qualitatively why this point on the ladder falls at an increasing rate of speed as the base is moved outward at a constant rate of speed. One can see that for each unit change in the distance that the base of the ladder moves outward, the amount that point *a* drops increases; an indication that the point is falling at an increasing rate of speed. Should we expect the same results for the midpoint? For a point closer to the base of the ladder? Repeating the process, one can clearly see that the new points placed on the ladder, *b* and *c,* also appear to drop at increasing rates of speed (see fig. 18.5); however, it is also clear that the rates at which the two points drop differ. In short, it appears that as the point on the ladder moves closer to the base of the ladder, the rate at which it drops decreases.

A next logical step to pursue concerns the nature of the actual curves that have been traced for the different locations of the point on the ladder. A reasonable guess is that the midpoint traces out a quarter-circle, whereas the

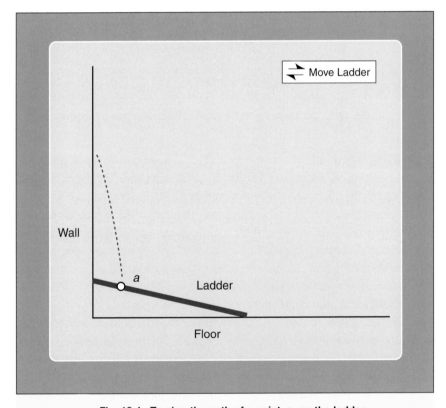

Fig. 18.4. Tracing the path of a point, *a*, on the ladder

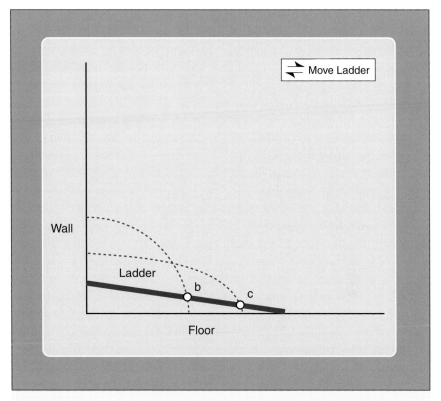

Fig. 18.5. Tracing the path of the midpoint, and a point below the midpoint

other two points trace out quarter-ellipses. Can we prove it? Similar to the TIMSS example discussed earlier, an important next step is to actually engage in proving these conjectures. Again, the use of technology suggests that these conjectures are reasonable, and the proof verifies that they are indeed true (in this instance) as well as offers insight into why the conjecture is true.

Considerations and Closing Remarks

Technological representations played a significant role in the solution process for each of the two problems, the TIMSS problem and the ladder problem. With respect to both tasks, technology allowed us to better represent (or model) the problem situation itself. In the first example, it allowed us to manipulate the vertices of the triangles, allowing us to see that for a given base, the areas remained the same. Such exploration also offers insight regarding the potential mechanism by which one might prove why the areas remained the same—the height of each triangle is the same. Students may have realized this fact visually from seeing the lengths of the important parts of the triangle themselves, or they may have also created an equation for the areas of the triangles and witnessed firsthand that when they manipulated the vertices of the triangles, the result of the formulas for the areas remained the same. In the second example, technology again allowed us to model the problem situation for the ladder, although this modeling was much more dynamic—and powerful—since it allowed us to literally animate the problem so that the ladder actually slid down the wall, and students could make mathematical observations or conjectures as the problem situation took place in our technological model. In a similar fashion, technology allowed us to continue modeling the ladder problem as we explored the extensions presented by the task.

In the five roles for technology use that we described in the beginning of this article, the modeling of the mathematics tasks also served as a *management tool* in helping both students and teachers more efficiently model the mathematics problem-solving situation (note that the "creator" of the ladder task, Steve Monk, actually brought a ladder into his calculus class and had students slide it down one of the classroom walls in order to model this problem). As we worked with technology to model both of these tasks with students, pre-service teachers, and in-service teachers, we have also observed how technology served as a *motivational tool* with these problems. Indeed, in all these learning situations, the challenge and utility of using technology to help solve these problems has often served as a catalyst to pull participants into engaging with the important mathematical ideas present in these tasks. Moreover, participants have often been motivated by the power of technology to model these tasks so that they have continued to explore these problems past the requirements of the task themselves. And perhaps most important, from a mathematics instructional point of view, technology served as a *cognitive tool* to help students better understand these mathematically rich tasks. As mentioned previously, technology first allowed students to understand these tasks better—the first process discussed in the introduction of this article—by modeling the problem-solving situation inherent in the context of each task.

In addition, technology allowed for a second cognitive process (discussed in the introduction) to take place—also, as was discussed in the introduction, this second process was set in the context of the first cognitive process of modeling the problem-solving situation. In particular, this second cognitive process focused on students' use of technology to explore a variety of mathematics procedures and concepts—required by the tasks—from a conceptual perspective. As was made clear in the discussion of the two mathematics tasks, this second process of using technology to better understand the mathematical concepts—of speed and rates, in particular—was more aptly fostered in the representation of the ladder problem, since students were able to harness technology to explore the mathematical ideas above and beyond the initial problem situation. In this fashion, technology served to represent the complexity of mathematical ideas so that both students and teachers more fully understood the conceptual underpinnings of the mathematics content presented in these rich tasks.

In closing, we wish to note that it is essential for teachers to integrate technology into their mathematics instruction and assessment as society moves forward in the "information age." Equally essential, however, is that we use technology thoughtfully in meaningful ways, so that we are not merely implementing technology for the sake of technology itself. In our work with mathematics students, preservice teachers, and in-service teachers, we have found these five roles for technology use—as a *management, communication, evaluation, motivation,* and *cognitive tool*—to be a helpful framework in guiding us in planning and implementing technology-based mathematics instruction and assessment in meaningful ways. Furthermore, as teachers and students become more familiar and comfortable with using technology in the mathematics classroom, they should strive to use technology as a cognitive tool, focusing on the two processes of representations and conceptual understanding, so that mathematics can be understood more fully and at a deeper level in order to allow students in grades K–12 to value and use the power of mathematics in their everyday lives.

REFERENCES

International Society for Technology in Education (ISTE). *National Education Technology Standards for Students—Connecting Curriculum and Technology.* Washington, D.C.: ISTE, 2000.

Knuth, Eric. "Fostering Mathematical Curiosity." *Mathematics Teacher* 95 (February 2002): 126–30.

Knuth, Eric, and Blake Peterson. "Fostering Mathematical Curiosity: Highlighting the Mathematics." *Mathematics Teacher* 96 (November 2003): 574–79.

Monk, Steve. "Students' Understanding of a Function Given by a Physical Model." In *Concept of a Function: Aspects of Epistemology and Pedagogy,* edited by Ed Dubinsky and Guershon Harel, pp. 175–93. Washington, D.C.: Mathematical Association of America, 1992.

National Council for the Social Studies (NCSS). *National Standards for Social Studies Teachers.* Silver Spring, Md.: NCSS, 2000.

National Council of Teachers of English (NCTE). *Standards for the English Language Arts.* Urbana, Ill.: NCTE, 1996.

National Council of Teachers of Mathematics (NCTM). *Curriculum and Evaluation Standards for School Mathematics.* Reston, Va.: NCTM, 1989.

———. *Professional Standards for Teaching Mathematics.* Reston, Va.: NCTM, 1991.

———. *Assessment Standards for School Mathematics.* Reston, Va.: NCTM, 1995.

———. *Principles and Standards for School Mathematics.* Reston, Va.: NCTM, 2000.

National Education Goals Panel (NEGP). *The National Educational Goals Report: Building a Nation of Learners.* Washington, D.C.: NEGP, 1991.

National Research Council (NRC). *National Science Education Standards: An Enhanced Sampler.* Working paper of the National Committee on Science Education Standards and Assessment. Washington, D.C.: NRC, 1993.

Perkins, David, Judah Schwartz, Mary West, and Martha Wiske. "Introduction." In *Software Goes to School,* edited by David Perkins, Judah Schwartz, Mary West, and Martha Wiske, pp. xiii–xvi. New York: Oxford University Press, 1995.

Thomas, George, and Ross Finney. *Calculus and Analytic Geometry.* 5th ed. Reading, Mass.: Addison-Wesley Publishing Co., 1979.

19

Using Technology to Show Prospective Teachers the "Power of Many Points"

Rose Mary Zbiek

THERE are many activities in which we might engage prospective teachers so they see technology as a means to enhance mathematical learning and teaching. It seems important to structure these experiences so each event helps the prospective teachers to achieve other goals, including deepening their understandings of mathematics, using other available resources well, and being ready to meet numerous demands placed on today's classroom teacher. This article describes a lesson sequence designed to make these many points with our prospective teachers. Combined with other similar lesson sequences, this experience helps emerging technology-savvy mathematics teachers to connect several aspects of their current college and future professional worlds.

The Prospective Teachers

The prospective teachers in our technology-focused methods course intend to teach mathematics in grades 7 through 12. They have earned at least eighteen credits of college mathematics and completed one early field experience. Many of them have been recently or are concurrently enrolled in a college-level, proof-rich geometry course. The group prides itself on its ability to solve typical mathematics problems that result in numerical answers. However, like many of their peers across the continent, they struggle to construct high-quality proofs and to see clearly how the college mathematics they are learning connects to the school mathematics they will be teaching.

By the time we get to this lesson sequence in roughly the sixth week of the fifteen-week course, our prospective teachers have had collective experience with graphing calculators and symbolic calculators and at least minimal expo-

Editor's note: The CD accompanying this yearbook contains a Geometer's Sketchpad file that is relevant to this article. For more information on The Geometer's Sketchpad, contact Key Curriculum Press at www.keypress.com/.

sure to The Geometer's Sketchpad (Jackiw 2001). The sequence begins in class and continues with assignments through the next week.

The Power of a Point Lesson Sequence

We begin with the task in figure 19.1, which can be solved using a common theorem in high school or college geometry. Given three to five minutes, some students conclude correctly that c is approximately equal to 1.94, others groan they should know how to "do it," and a few give up in frustration. At least one student notices that the task relates to the power of a point theorem from his or her college geometry course. This comment inspires other students to recognize this connection or to inquire what "power of a point" means.

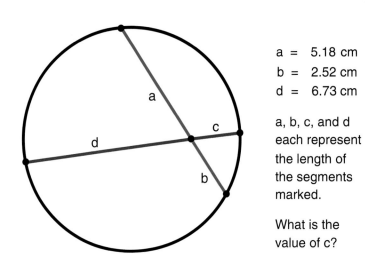

a = 5.18 cm

b = 2.52 cm

d = 6.73 cm

a, b, c, and d each represent the length of the segments marked.

What is the value of c?

Fig. 19.1. The initial problem

We then consider ways in which secondary school students who are given this problem might use technology. The clear majority of the prospective teachers observe that calculators of most any type could be used either to check the "arithmetic" done by hand or to compute initially the answer as $5.18 \times 2.52 \div 6.73$ (from $c = a \times b \div d$). A few students note that Sketchpad could also be used to check the computation and possibly to compute or at least establish the value of c. When asked about how to use Sketchpad to do this, they note they would have to determine how to get three segments with these exact measures, and they add the caveat that this method of finding the value of *c* would be extremely more difficult than the calculator method. As one student commented, "Why would we *bother* doing this?" This brings home the point that technology use for the sake of technology use is not our goal.

Perhaps unsurprisingly, the prospective teachers are very sensitive to the fact that some of them could *not* solve this "easy" problem. However, we started the semester agreeing that admitting we did not know something and then learning about that thing would always be better than pretending we know more than we do know. We confirm that the observed inability of some students to solve this problem is not surprising when we realize the problem is very "easy" *only if* the students already "know" the relationship among the four segment lengths—a relationship that many of us did not recognize or did not connect to the problem until we were reminded about it. Our bigger question as prospective teachers then is, assuming we value students' knowing about this relationship, how do we get our students as a group to know the relationship *better* than we as a group knew it?

This question raises for the prospective teachers an issue of when secondary school students might first encounter the relationship. The prospective teachers want to look for the "formula" in textbooks—many of the students initially believe textbooks should furnish all the mathematical experience their students need and all the mathematical background information they as teachers will ever need. We turn to (premarked) pages in each of a dozen textbooks where high school geometry students are introduced to the theorems related to the power of a point. The textbooks include materials currently in use in the schools the students attended or visited (e.g., Sallee et al. 2000) as well as books in use while these students were in high school (e.g., Serra 1993), and books that predated the lives of many of the students (e.g., Rhoad, Milauskas, and Whipple 1991; Hirsch et al.1979; Sobel et al.1990; Coxford and Usiskin 1975; Nichols et al. 1986; Keniston and Tully 1960; Rising et al. 1981). This gives the group a sense of how the presentation of secondary school content in textbooks may vary over time and across different books within the same era. We note differences among the textbooks (e.g., expressing the relationship as equal products of measures of segments versus a proportion, the degree of formality of any proofs offered, the difficulty and diversity of exercises or problems), yet recognize that many books tend to present a set of three related theorems, a version of which appears in figure 19.2.

In small groups the prospective teachers discuss how these three theorems in figure 19.2 are similar to, and different from, each other. The prospective

Chord-Chord Power Theorem

If two chords of a circle intersect inside the circle, then the product of the measures of the segments of one chord is equal to the product of the measures of the segments of the other chord.

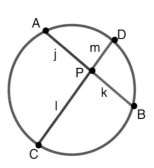

$$j \cdot k = 16.51 \text{ cm}^2$$
$$l \cdot m = 16.51 \text{ cm}^2$$

Tangent-Secant Power Theorem

If a tangent segment and a secant segment are drawn from an external point to a circle, then the square of the measure of the tangent segment is equal to the product of the measures of the entire secant segment and its external part.

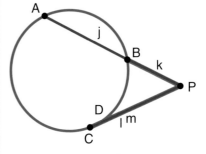

$$j \cdot k = 51.66 \text{ cm}^2$$
$$l \cdot m = 51.66 \text{ cm}^2$$

Secant-Secant Power Theorem

If two secants are drawn from an external point to a circle, then the product of the measures of one secant segment and its external part is equal to the product of the measures of the other secant segment and its external part.

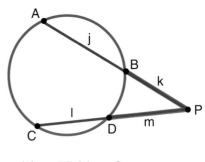

$$j \cdot k = 75.64 \text{ cm}^2$$
$$l \cdot m = 75.64 \text{ cm}^2$$

Fig. 19.2. Typical statements of the three theorems (summarized from Rhoad, Milauskas, and Whipple 1991, pp. 493–94)

teachers note the presence in all three theorems of four segment lengths, the equating of two products of segment lengths, and the importance of "some point of intersection." The groups' next task is to brainstorm how they can use Geometer's Sketchpad (GSP) to help high school students see these mathematical connections. Most groups focus on having one GSP file to illustrate each of the three theorems.

Things change as our whole-group discussion points out that most of the groups use the three GSP files as merely computer-based static images of what teachers or students could easily draw on paper, whiteboard, or transparency. The class is challenged to consider how the *dynamic* nature of Geometer's Sketchpad may be used to get at the essential connections among the three theorems. Eventually someone notices the point of intersection is the thing they want to see "move," and we set out to create a computer file that allows the user to drag the intersection point. The goal is to use this file to show the measures and produce examples as the point of intersection is dragged inside and outside the circle. This is not an obvious GSP construction for the prospective teachers—particularly for those relatively unfamiliar with the technology. The prospective teachers initially suggest a strategy that yields the picture in figure 19.1: construct a circle, construct two chords, construct the point of intersection of the two chords, measure four resulting segments, calculate c, and be done. My students have a sense of accomplishment in that they learned enough about GSP to carry out their ideas confidently and to create quickly a figure that successfully represents one of the theorems, but they realize the figure cannot be used to represent the other theorems. The prospective teachers try alternative strategies to construct one sketch to represent all three theorems, and we eventually arrive at a sketch that is sufficiently robust. A minimally robust sketch in this situation is one in which, as the student drags the intersection point, the intersection point continues to exist, each of the chords may become a secant segment or a tangent segment, and the measures for a, b, c, and d remain numerically correct for relevant locations of the intersection point. As a safeguard, I have a prepared file ready.[1] Figure 19.3 shows the result of dragging the intersection point in a robust sketch to create several cases.

My students also are expected to understand the proofs of theorems. Instead of being asked to "study the proofs" as their assignment, the prospective teachers consider the three theorems and proofs within the context of a teaching question. For example, students have the GSP file illustrated in figure 19.3 as well as access to a JavaSketchpad applet for the next part of the

---

1. A copy of this file is in the CD accompanying this yearbook. The file name is Chord/Secant/Tangent2.gsp.

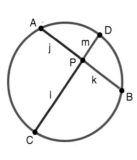

j = 4.15 cm   j·k = 15.58 cm²
m = 2.34 cm   l·m = 15.58 cm²
k = 3.75 cm
l = 6.66 cm

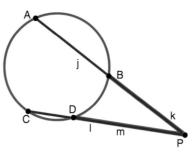

j = 16.41 cm   j·k = 138.67 cm²
m = 10.12 cm   l·m = 138.67 cm²
k = 8.45 cm
l = 13.71 cm

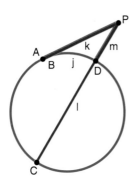

j = 6.62 cm   j·k = 43.61 cm²
m = 3.49 cm   l·m = 43.61 cm²
k = 6.59 cm
l = 12.50 cm

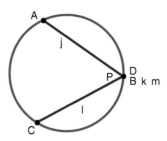

j = 7.73 cm   j·k = 0.14 cm²
m = 0.02 cm   l·m = 0.14 cm²
k = 0.02 cm
l = 7.72 cm

Fig. 19.3. GSP figures show results of dragging the intersection point.

assignment, which asks them to develop two lesson ideas. (A lesson *idea* is a description of the general flow of a lesson but with far less detail than a lesson *plan*; the purpose of developing lesson ideas is to encourage the prospective teachers to focus on major learning goals during the initial planning stage of a lesson.) The goal of one of the two lessons is to have secondary school students know and be able to use a "formula" to solve problems similar to the problem in figure 19.1. The second lesson should use the file or applet to help students understand the proofs of the theorems. A sample lesson idea for each of the two goals appears in figure 19.4. For prospective teachers to do this task well, they likely need time outside of a class meeting. My students get individual feedback on their ideas, and we return as a group to discuss how the mathematical learning goal of a lesson should guide their instructional choices, including the technology use in the lesson. For example, given the idea for a lesson about understanding the proofs as shown in the right-hand column of figure 19.4, students note that this lesson idea relies on demonstration by the teacher. The prospective teachers then debate whether the lesson could be more effective if the high school students working in pairs used the software directly instead of relying on their teacher's demonstration. We consider how, if at all, the teacher might guide students through the reasoning that underlies the proof suggested in figure 19.4. Our discussion then turns to how we can use the dynamic nature of the file to address the secant-secant and tangent-secant cases in ways that help secondary school students to understand the theorems and their proofs (e.g., why a product of segment lengths is common to the theorems, what the role of similar triangles is in the proofs).

Beyond Lesson Development

Several related discussions and assignments require the prospective teachers to go beyond creating and critiquing lesson ideas. In an obvious component of the follow-up work, the prospective teachers discuss how their college mathematics content arises in the high school textbooks. My students are fascinated to see how the power-of-a-point ideas could become the essence of three theorems in a secondary school mathematics textbook. We look in the high school geometry books for other examples of how college mathematics may appear explicitly and implicitly in secondary school materials, and how Geometer's Sketchpad may be useful in understanding these examples. This detective work helps prospective teachers look beyond the mathematics on the secondary school textbook pages and draw on their college mathematics to see connections, to create examples, and to challenge students. This work also relieves some of the college students' anxiety about spending so many hours in mathematics courses when what they really want to do is "get out and teach."

The prospective teachers also connect the lessons they devise to National Council of Teachers of Mathematics (NCTM) and state standards at different grade levels. One list of possibilities appears in figure 19.5. Beyond simply generating these lists, we talk about how the standards addressed involve both content issues (e.g., knowing particular properties in 2.9.11.F [of the

Students will use the chord-chord theorem to solve problems.	**Students will understand a proof of the chord-chord theorem.**

Use the sketch (replicated here) to generate several cases (as in fig. 19.3).

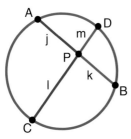

$$j \cdot k = 16.51 \text{ cm}^2$$
$$l \cdot m = 16.51 \text{ cm}^2$$

Notice the two products are always equal and express this equality as an equation.

Students working in pairs write in words and in symbols what this equation means in the figures. (Students thus state the theorems as well as internalize verbal versions of the equation.)

Discuss suggested written responses, including alternative correct forms and revising incorrect versions.

Practice applying the formula by solving exercises similar to the task in fig. 19.1, using calculators as needed.

Show a circle and two intersecting chords and the equation jk = ml.

Students brainstorm a list of what they "know" about the figure, equation, and setting.

The teacher draws from the list things related to the congruence of vertical angles and to how the equation could arise from a proportion if $\triangle PAC \sim \triangle PDB$ (given figure below). The teacher asks students to prove that these triangles are similar.

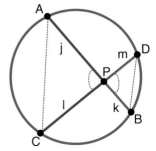

If students struggle to establish triangle similarity, draw attention to minor arc *CB* and ∠A and ∠D subtend this arc to get ∠A ≅ ∠D. Use Sketchpad to make arcs more apparent (see figure below).

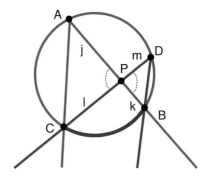

Drag *P* within the circle to show that the existence of vertical angles and the subtended *BC* arc exist for all points interior to the circle. Note how this generality is needed in order to show the chord-chord theorem.

Fig. 19.4. Prospective teachers' early suggestions as lesson ideas

Pennsylvania state standards]) and process issues (e.g., conjecturing in general in 2.4.8.A). Then, we compare how the use of technology to address content issues may differ from the use of technology to address process issues. The prospective teachers see how they can devise lessons that address multiple standards as well as use technology in concert with external demands placed on their future schools and districts.

Grade 8	Grade 11
2.4.8.A. Make conjectures based on logical reasoning and test conjectures using counterexamples. 2.1.8.D. Apply ratio and proportion to mathematical problem situations involving distance, rate, time, and similar triangles.	2.9.11.B. Prove that two triangles or two polygons are congruent or similar using algebraic, coordinate, and deductive proofs. 2.9.11.F. Use the properties of angles, arcs, chords, tangents, and secants to solve problems involving circles.

Fig. 19.5. Sample of relevant Pennsylvania state standards

A third extension asks the prospective teachers to link their lesson ideas to mathematics education research. Prior to this lesson sequence, my students read Glass and Deckert's (2001) discussion of research involving interactive geometry. The prospective teachers then write a reflection about how their lesson ideas address principal points in that article as well as how they could refine their lesson ideas to embody Glass and Deckert's research-based observations. Some typical observations include the following points, with important ideas from Glass and Deckert's article in italics:

- Differences between Sketchpad as an *interactive* construction environment rather than a *static* construction environment like The Geometric Supposer (Schwartz and Yerushalmy 1985) or paper-and-pencil work. For example, using the interactive diagram allows students to drag a point and quickly generate many examples for one theorem and to see what is similar across the three different theorems.

- The importance of *focusing on relevant parts* of the diagram and relationships among these parts in order to apply the formula as well as to understand claims in the proof. For example, looking at the highlighted arc and the angles that subtended it rather than the intersection point or chords was important in understanding the proof of the chord-chord case, and good use of the technology can help students focus on these components of the figure.

- The importance of students having *a construction versus a drawing* to be sure the necessary relationships among points, segments, and circle remain intact. For example, for the tangent-secant case, several prospective teachers initially placed a line on the screen that appeared tangent to the circle until they dragged a point—they had created a *drawing* rather than a *construction*. A construction is created in mathematically correct ways, and it retains its mathematical properties during dragging; a drawing appears to be mathematically correct but fails to retain its mathematical properties during dragging. In this tangent-secant example, when the students drag the intersection point, the line in the drawing is no longer tangent to the circle.

This reflection serves both as an assessment of their understanding of the *Mathematics Teacher* article (Glass and Deckert 2001) and as a means by which the prospective teachers become accustomed to making research-based classroom decisions.

Closing

The power-of-a-point work is one of several lesson sequences in the technology-focused methods course discussed here. Developing these sequences usually started with ideas from a research-for-the-classroom article (typically drawn from connecting-research-to-practice sections of NCTM school journals) about a form of technology available in schools. Building on problems or activities from those sources leads to ideas that help to draw mathematical connections between students' college courses and the secondary school curriculum through lessons that address an explicit set of state and national standards. For example, another lesson sequence involves using graphing calculator–based data-gathering tools with corresponding journal articles (Beckmann and Rozanski 1999; Doerr, Rieff, and Tabor 1999) to connect function and calculus concepts from college courses to several Pennsylvania mathematics standards for grades 8 and 11.

In each of the lesson sequences, the prospective teachers use technology to solve problems, to develop mathematical theory, and to enrich their own conceptions. Each lesson sequence connects to state standards, college mathematics experiences, and research insights. During the semester, the extent to which the prospective teachers (or their [hypothetical] students) create their own files in Sketchpad or in other environments varies greatly. The perceived roles of technology also vary extensively, and the prospective teachers move from seeing technology only as a way to check answers to seeing technology as a venue for illustrating mathematical principles, for exemplifying different aspects of mathematical concepts, and for illuminating mathematical connections. These emerging technology-savvy teachers move from seeing technology as "another thing to teach" or "another thing to learn" to a tool to help students of all ages develop deep understanding and to meet the professional expectations that await them in their future classrooms.

REFERENCES

Beckmann, Charlene E., and Kara Rozanski. "Graphs in Real Time." *Mathematics Teaching in the Middle School* 5 (October 1999): 92–99.

Coxford, Arthur F., and Zalman P. Usiskin. *Geometry: A Transformation Approach.* River Forest, Ill.: Laidlaw Brothers, 1975.

Doerr, Helen M., Cathieann Rieff, and Jason Tabor. "Putting Math in Motion with Calculator-Based Labs." *Mathematics Teaching in the Middle School* 4 (March 1999): 364–67.

Glass, Brad, and Walter Deckert. "Making Better Use of Computer Tools in Geometry." *Mathematics Teacher* 94 (March 2001): 224–29.

Hirsch, Christian R., M. A. Roberts, Dwight O. Coblentz, Andrew J. Samide, and Harold L. Schoen. *Geometry.* Glenview, Ill.: Scott Foresman, 1979.

Jackiw, Nickolas. The Geometer's Sketchpad. Ver. 4.0. Software. Emeryville, Calif.: Key Curriculum Press, 2001.

Keniston, Rachel P., and Jean Tully. *High School Geometry.* Boston: Ginn, 1960.

Nichols, Eugene Douglas, M. L. Edwards, E. H. Garland, S. A. Hoffman, A. Mamary, and W. F. Palmer. *Holt Geometry.* New York: Holt, Rinehart & Winston, 1986.

Rhoad, Richard, George Milauskas, and Robert Whipple. *Geometry for Enjoyment and Challenge.* Evanston, Ill.: McDougal Littell, 1991.

Rising, Gerald R., John A. Graham, John G. Balzano, Janet M. Burt, and Alice M. King. *Houghton Mifflin Unified Mathematics* (Book 3). Boston: Houghton Mifflin, 1981.

Sallee, Tom, Judy Kysh, Elaine Kasimatis, and Brian Hoey. *College Preparatory Mathematics* 2 (Geometry). 2nd ed. Sacramento, Calif.: CPM Corp., 2000.

Schwartz, Judah L., and Michal Yerushalmy. The Geometric Supposers. Software. Pleasantville, N.Y.: Sunburst Communications, 1985.

Serra, Michael. *Discovering Geometry.* Berkeley, Calif.: Key Curriculum Press, 1993.

Sobel, Max A., Evan M. Maletsky, Neal Golden, Norbert Lerner, and Louis S. Cohen. *Geometry.* Mission Hills, Calif.: Glencoe/Macmillan, 1990.

20

Investigating Mathematics with Technology: Lesson Structures That Encourage a Range of Methods and Solutions

Rebecca McGraw
Maureen Grant

PLANNING and implementing classroom activities in ways that maximize opportunities for learning has always been a central activity of teaching. Technology is an important tool in this process because it offers unparalleled opportunities for investigating and understanding mathematics. Of course, these benefits are not realized merely by bringing technology into the classroom; lessons must be structured in ways that encourage students to develop their problem-solving and reasoning skills, make connections among concepts and representations, and communicate their ideas to one another and to their teachers. Lessons must also be structured in ways that allow teachers opportunities to listen and respond to their students' ideas and assess their understanding as they work.

In this article, we consider how the structure of technology-based lessons influences students' opportunities to learn mathematics as described in *Principles and Standards for School Mathematics* (National Council of Teachers of Mathematics [NTCM] 2000). Specifically, we compare two kinds of lessons (we call these Type 1 and Type 2) according to the extent to which they engage students in (1) identifying patterns and searching for relationships, (2) making and investigating mathematical conjectures, and (3) developing and evaluating mathematical arguments. In addition, we discuss the kind of students' work produced by the two lesson types and the potential of this work to become the basis for an analysis and evaluation of mathematical thinking. The examples we furnish throughout the article are ones that we have some experience using with algebra and geometry students. Although

---

Editor's note: The CD accompanying this yearbook contains a Word document, "Examples of Type 1 and Type 2 Technology-Based Lessons," (under the name Examples.doc) that is relevant to this article.

the examples we use are taken from the secondary school level, we believe that the ideas we discuss are equally useful when designing activities for younger students.

Technology-Based Lessons—Type 1

In lessons of this type, all students proceed through the lesson in the same way and, if all goes well, reach the same conclusions. Instructions are written to focus students' attention on particular mathematical relationships. Technology is used to help make these relationships transparent to students (fig. 20.1).

For the purposes of this discussion, we focus not on the particular equations or shape chosen in the examples (see fig. 20.1) or on the wording of the directions but rather on the overall structure of the activities. Specifically, in technology-based lessons of this type—

Use your graphing calculator to compare the graphs of the following equations.

$$y = x^2$$
$$y = \frac{1}{2}x^2$$
$$y = 2x^2$$
$$y = 5x^2$$
$$y = -x^2$$
$$y = -\frac{1}{2}x^2$$
$$y = -2x^2$$
$$y = -5x^2$$

Describe the effect of changing the value of *a* on the graph of $y = ax^2$.

[Students are given instructions for constructing a parallelogram using interactive geometry software, or are provided with a parallelogram on a disk or through file-sharing.]

Measure the lengths of the sides of the parallelogram.
[Students are given instructions for measuring length.]

Measure each angle of the parallelogram.
[Students are given instructions for measuring angles.]

Study the side lengths and opposite and adjacent angle measures as you change the size and shape of the parallelogram. Make several conjectures about the relationships among side lengths and among angle measures of parallelograms.

Fig. 20.1. Examples of technology–based lessons—Type 1

- the instructions require that students identify relationships and make conjectures;

- the instructions specify both what is to be investigated and how it is to be investigated;

- students move sequentially through the instructions with little or no decision-making required;

- if the lesson is implemented successfully, students will have used similar methods and made similar conjectures.

In the graphing example in figure 20.1, students use their calculators to study a particular set of functions. The functions are chosen to highlight the effect of changing the value of a specific parameter, in this example a, on the graph of the function. One can imagine similar sets of equations that could be given to students to help them determine the effect of varying b or c in $y = ax^2 + bx + c$. Technology-based lessons like this one are often used with linear, exponential, logarithmic, and trigonometric functions as well. Although not included in the example, we can imagine teachers following up this lesson by asking students to explain why changing the value of a affects the graph as it does, as was done at the end of the following classroom vignette (fig. 20.2). This would be one way to engage students in attempting to prove a conjecture.

Mr. Jeffries stands in front of his class of thirty ninth-grade algebra 1 students. Students' desks are arranged in groups of three or four. Mr. Jeffries writes eight equations (see fig. 20.1) on the overhead projector.

Mr. Jeffries: Please copy these equations in your notebooks. We're going to use our graphing calculators today to graph and compare these equations.

[Students copy down the equations.]

Mr. Jeffries: Notice that all these equations are of the form $y = ax^2$ [*he writes "$y = ax^2$" on the overhead projector*], but the value of a is different in each equation. For example, in the second equation, a is $\frac{1}{2}$ and in the third equation, a is 2. What is the value of a in the first equation?

[Some students say, "One."]

Mr. Jeffries: One. That's right. Remember when there's no coefficient written in front of the variable, that means the coefficient is 1. Now, as you are graphing these equations, I want you to be thinking about how changing the value of a affects the graph.

[Mr. Jeffries writes, "Describe the effect of changing the value of a on the graph of $y = ax^2$" on the overhead projector.]

Mr. Jeffries: I would like you to work with the people in your group on this activity, but I want each of you to write down your findings in your notebook. Also, you may not be able to put all eight equations on one graph and still be able to see which is which, so you may want to make some sketches of the graphs in your notebooks as you are working. Remember that you can turn the equations on and off, so you don't have to look at all the graphs at once. Oh, I forgot to tell you what viewing window to use. Let's use the standard viewing window for this activity.

[Students take out their calculators and begin to enter the equations. Mr. Jeffries moves from group to group, reminding students how to turn equations on and off and reminding them how to set their viewing windows. Students spend approximately fifteen minutes graphing the equations, sharing ideas with one another, and writing in their notebooks.]

Mr. Jeffries: I would like each group to share some of their findings with the rest of the class. Who will volunteer to go first? Marta?

Marta: If the equation has a negative in it, then the graph goes down.

Mr. Jeffries: A negative where, exactly?

Marta: In front of the number, the *a*.

Mr. Jeffries: Did other groups find this? *[Many students nod.]* Did you say "goes down" or did you say it differently? Peter?

Peter: I said, "It points down."

Mr. Jeffries: OK. Ruben?

Ruben: We said that if there's a negative, it flips the graph. Like if it was plus $2x$, then it would go up, but if it was negative $2x$, then it would be the same, just upside down.

Mr. Jeffries: OK, let me try to sum up what each of you has said.

[Mr. Jeffries writes on the overhead projector, "If the value of a is negative, then the graph opens down. Changing the a value from positive to negative causes the graph to 'flip.'"]

Mr. Jeffries: Do you agree with that? *[Many students nod or say yes.]* OK. What other things did you find? Sarah?

Sarah: I said that the bigger the number is, then the narrower the graph is.

Mr. Jeffries: Did anyone else have that? Yes?

Jamie: We had that, too.

Adam: We had the smaller the number, the flatter it is.

[Several other students say, "We had that."]

| Mr. Jeffries: | Did you find that with the negative numbers as well? |

[Several students say yes.]

Mr. Jeffries:	Which is narrower, negative five or negative two?
Students:	Negative five.
Mr. Jeffries:	But negative five is *smaller* than negative two.
Tina:	Yeah, but it's a bigger negative number, so it's narrower.
Mr. Jeffries:	The number itself is smaller, but the absolute value is larger. Let's write it this way. *[Mr. Jeffries writes as he talks.]* The larger the absolute value of *a* is, the narrower, or less open, the graph is. The smaller the absolute value of *a* is, the wider, or more open, the graph is.
Mr. Jeffries:	Did anyone find anything else? Nate?
Nate:	They're all curves.
Mr. Jeffries:	Curved how?
Nate:	They're all the same shape, they're not straight lines.
Mr. Jeffries:	Did anyone else have that? Anna?
Anna:	We had, well, because of the squared, the exponent, that makes them all curve up or down.

[Mr. Jeffries writes, "The equations are quadratic, and the graphs are parabolas."]

| Mr. Jeffries: | *Quadratic* means that the highest exponent on the variable is 2. Have you heard the word *parabola* before? *[A few students nod.]* The curved shape these graphs all have is called a parabola. OK. Did anyone have anything else? *[Mr. Jeffries waits.]* No? OK. The next thing I want you to think about is *why* these equations have the shapes they have. Your group should choose two of the equations, whichever ones you want, and create tables of values for each equation. The *x*-values for each table should be these: –10, –5, –4, –3, –2, –1, 0, 1, 2, 3, 4, 5, 10. *[Mr. Jeffries writes these numbers on the overhead projector.]* You can calculate the *y*-values for the tables. When you plug these *x*-values into the equations, think about how the operations, the exponents, and the multiplication are affecting the *x*-values. The goal is for you to be able to explain *why* each graph has the shape it has. *Why*, when you have an exponent, is the graph curved? *Why* does the graph change when the numbers in front of the variable change? |

[The students begin creating tables.]

Fig. 20.2. Classroom vignette of a Type 1 technology-based lesson

As is typical of Type 1 technology-based lessons, students in this vignette generated a small number of similar conjectures. Because the equations were a convenient subset of the quadratics, the effects of changing the value of a on the graphs were readily apparent to students. The lesson did not require students to make decisions about what to do or how to do it. Initially, students were not asked to prove their conjectures; however, as a second part to the lesson, Mr. Jeffries asked students to create tables and consider why changing the value of a affects the graph as it does.

Instructions for the second example in figure 20.1 also specify what is to be investigated (relationships among the lengths and angles of a parallelogram) and how it is to be investigated (by measuring and comparing measurements). In addition, students are either given a figure (parallelogram) to manipulate or step-by-step instructions to create one. The parallelogram is a relatively simple figure to construct; however, more complex figures and relationships could be the focus of a lesson of this type. For example, students could construct medians, altitudes, and perpendicular bisectors of the sides of a triangle and investigate the relationship among the centroid, orthocenter, and circumcenter (the points formed by the intersections of the medians, altitudes, and perpendicular bisectors, respectively). As with the equations example, students might also be asked to prove the conjectures they make.

Relating Type 1 lessons to learning mathematics as described in *Principles and Standards for School Mathematics* (NCTM 2000), we can see that lessons of this type may be useful for presenting students with opportunities to make connections among mathematical ideas and representations and to recognize and describe relationships within and among objects. In addition, teachers can follow up these kinds of activities by asking students why certain relationships seem to exist as a way to engage them in developing mathematical arguments. Type 1 lessons give students only limited opportunities, however, to identify patterns and search for relationships, investigate conjectures, and reason mathematically. Students can move through much of these lessons without reasoning, reflecting, and making decisions.

When we began implementing technology-based lessons in our own classrooms, the lessons we designed were mostly Type 1. We became dissatisfied, however, when we observed students spending large portions of their class time attempting to understand and follow our carefully written directions rather than thinking deeply about the mathematics they were investigating. We were equally frustrated by the amount of time we spent during class telling students what to do and how to do it when our hope had been that the technology would help students become the investigators. From a practical standpoint, we were often not able to work our way around the room and respond to students' questions in a timely manner; this led to off-task behavior and classroom management problems when students believed that they could not move forward in their work until their questions had been answered. On the positive side, waiting for help caused some students to find ways to move forward on their own. We noticed that these students sometimes started doing the kinds of investigating we had hoped for, that is, looking for relationships among shapes, numbers, and representations and using the technology as a tool for uncovering patterns and connecting ideas.

We began to wonder whether we could redesign our technology-based lessons in ways that would incorporate more opportunities for students to make decisions. We wanted to encourage reasoning and reflection and to create a classroom environment in which students spent less time following our directions and more time investigating and developing their own ideas. From a practical standpoint, we needed to find ways to encourage students to work independently from us, so that all students could be productively engaged throughout the class period. We began to consider ways to allow for independence and variation in students' work while at the same time fostering meaningful engagement with important mathematics. Characteristics of these new technology-based lessons, which we call Type 2, are discussed in the following section.

Technology-Based Lessons—Type 2

Type 2 lessons (fig. 20.3) are similar to Type 1 lessons in that technology is used as a tool for investigating mathematical relationships. Type 2 lessons differ from Type 1 in both the nature of the instructions and the roles of the teacher and students during implementation. In Type 1 lessons, the teacher seeks to *simplify* the investigation by choosing a particular object or set of objects and a method of investigation that leads students to notice a particular mathematical relationship. In Type 2 lessons, the teacher seeks to *complexify* the investigation by choosing a range of objects (in the examples presented in fig. 20.3 the objects are graphs and quadrilaterals) in which are embedded multiple and overlapping relationships. *Students* do the simplifying as part of the investigation. Because a range of objects is offered, students may notice many relationships but choose only a few for closer examination. The ways in which these closer examinations occur are partly determined by the teacher's instructions, but room is created for students to make decisions as well.

To summarize, Type 2 lessons have the following characteristics:

- The instructions require that students identify relationships and make conjectures.
- Students play a role in determining what to investigate and how to investigate it.

[Students are given the graphs (but not the equations) for the following functions:]

$$y = -x \qquad y = -2x^2 \qquad y = x^2 - 4$$

$$y = 2x \qquad\qquad\qquad\quad y = 3x + 1$$

$$y = \frac{1}{2}x$$

$$y = x^2 \qquad\qquad\qquad\quad y = (x-3)^2$$

$$y = -3x^2$$

$$y = -3x \qquad\qquad\qquad y = 2x - 1$$

$$y = -2x$$

$$y = 4x \qquad\qquad\qquad\quad y = x - 3$$

$$y = -x^2$$

$$y = 2x + 4 \qquad y = 2x^2 + 1$$

$$y = 3(x-3)^2 \qquad y = -x^2 + 3$$

$$y = -\frac{1}{2}x^2 \qquad y = 2x + 2$$

$$y = x + 3 \qquad\quad y = (x+2)^2$$

$$y = x^2 - 2 \qquad\quad y = x + 4$$

Study the graphs. Notice their overall shapes, maximum and minimum values (if any), and *x*- and *y*-intercepts. Organize the graphs into groups and subgroups. Write a description of how you grouped the graphs.

[Students are given equations that match the graphs.] Which equation do you think matches which graph? Why?

Graph the equations with your graphing calculator to check your predictions. Write each equation beneath its graph.

What conjectures can you make about the relationships between the numbers and operations in the equations and the shape and location of the graphs? What equations (not on the list) could someone use to test your conjectures?

[Students have previously constructed and saved several different types of quadrilaterals using interactive geometry software or they have been given the quadrilaterals on a disk or through file sharing.]

Select a quadrilateral to examine.

Drag vertices and sides to change the size and shape of the quadrilateral.

What conjectures can you make about the properties of the quadrilateral? Feel free to construct midpoints, diagonals, and parallel and perpendicular lines; you could also measure lengths, angles, or areas.

Do your conjectures seem to hold for other quadrilaterals? Why or why not?

Fig. 20.3. Examples of technology-based lessons—Type 2

- Mathematical reasoning and decision making are required throughout the lesson.
- If the lesson is implemented successfully, students will have used a variety of methods and made a variety of conjectures.

Before describing how these characteristics are embodied in each of the Type 2 examples above, we need to make an important comment. When one compares Type 1 and Type 2 lessons, it is immediately apparent that Type 2 lessons would take much more time than Type 1 lessons to implement. We solve this problem by substituting one Type 2 lesson for several Type 1 lessons. For example, a series of Type 1 lessons could be used to identify the effect of changing parameters *a*, *b*, and *c* on the graphs of $y = ax + b$ and $y = ax^2 + bx + c$. We would argue that the content covered by a series of Type 1 lessons over several days could be covered in approximately the same amount of time using one Type 2 lesson. Of course, with the Type 2 lesson, every student would not investigate every parameter; however, individual students' findings could feed into a class pool of knowledge. The vignette in figure 20.4 is a synopsis of the implementation of a Type 2 lesson involving graphing calculators (see fig. 20.3 for a description of the lesson).

Day 1

[Ms. Martinez has twenty-eight algebra 1 students sitting in pairs. After a short introduction to the lesson, she gives students the twenty-five graphs and instructs them to organize the graphs into groups and subgroups and write explanations of how they organized their graphs. Students spend the rest of the class period analyzing and organizing the graphs and writing their descriptions. Ms. Martinez moves from table to table, asking students why they chose to put certain graphs in the same group. She does not offer students suggestions about which graphs should go together but rather encourages students to be explicit about the reasons for their decisions. Ms. Martinez's questions help her gather information about how her students are thinking about the characteristics of graphs.]

Day 2

Ms. Martinez:	What characteristics did you use to organize the graphs? Jessica?
Jessica:	Whether it was curved or straight.
Ms. Martinez :	*[Writes "curved or straight" on the board.]* Yes?
Antonio:	If it was a line, if it went down or up. *[Antonio moves his hand to show what he means by "down or up."]*

[Ms. Martinez writes "lines that decrease" and "lines that increase" with sketches of lines with negative and positive slopes beside each phrase, respectively. She continues writing students' ideas on the board.]

Ms. Martinez:	Now that you have heard the characteristics other people used to organize their graphs, take a few minutes to look at your own

work. Do you want to change any of your groups, or move a graph from one group to another?

[Students spend a few minutes reviewing their groupings.]

Ms. Martinez: Next, I am going to give you the equations that match the graphs.

[Ms. Martinez distributes the equations to the students.]

Ms. Martinez: Let's look at these equations. What are the components of these equations? What are they made up of?

[Students offer suggestions, including numbers, operations, addition, sub-traction, multiplication, division, exponents, parentheses, negative numbers, x's and y's, and equal signs. Ms. Martinez writes all these ideas on the board.]

Ms. Martinez: Now, all these things you have mentioned affect the graph of the equation in some way, and they are related to the character-istics of the graphs that you have already been thinking about. When you look at the equations, what do you notice that some of them have and others don't have?

[Students' ideas include parentheses, several operations, exponents, frac-tions, and negative signs in front of the x. Ms. Martinez writes all these ideas on the board.]

Ms. Martinez: Good. You are already noticing many characteristics of these equations. Now, keeping these characteristics in mind, I want you to work with your partner and make some educated guesses about which equations might go with which graphs. Write the equation beside the graph you think it goes with. Remember, I don't expect you to match all of them correctly, but I do expect you to use good reasoning.

[Students spend the rest of the class period matching the equations and graphs. Ms. Martinez moves from table to table listening to students as they discuss their ideas with one another.]

Day 3
[Ms. Martinez begins by asking the students to share a few ideas about which equations match which graphs and why. She then instructs her students to use their graphing calculators to graph the equations and check their predictions.]

Ms. Martinez: Now that you have checked your predictions, were there any graphs that surprised you? Which equations did you match cor-rectly? Michael?

Michael:	We knew that some of them were lines, so we got some of them right.
Ms. Martinez:	Some of them?
Michael:	Some of them we mixed up. It *was* a line, but it didn't go the way we thought it would, like up or down.
Ms. Martinez:	Eunice?
Eunice:	We thought if it had a negative, it would go down, but it depended where the negative was, like if it was by only the number, then it didn't go down.

[Ms. Martinez writes on the overhead projector: $y = x - 3$ and $y = -2x$.]

Eunice:	Yeah. The first one wouldn't go down, but the second one would.
Ms. Martinez:	Jon?
Jon:	We got that, too, but also, it works for exponents. The negative makes it go down.
Holly:	But only if it's by the *x*.
Jon:	Right.

[Students continue to share their ideas.]

Ms. Martinez:	Now it is time for you to write down the relationships you found. If you have some ideas you think might be correct, but you aren't sure, please write those down, too. Also, I want you to write your ideas down in if-then form, so that each idea looks like this. *[Ms. Martinez writes: "If the equation has _____, then the graph _____."]* You should use your graphing calculator to graph some equations and test your ideas. Underneath each of your if-then statements, write down a few equations that someone could graph that would illustrate your idea. Be sure that at least two of the equations are ones that are *not* on the list of twenty-five. Also, please make sure you clearly explain the terms you use. For example, what exactly do you mean when you say the graph "goes down"?

[Students begin this assignment in class and complete it at home. Ms. Martinez collects students' work and creates a class list of if-then statements, which she distributes to her students. She then asks each student to choose two of the statements, one that they think is true and one that they think might not be correct. She asks students to write an explanation of why the statement they believe is true must be true and why the statement they believe is false is false. For the false statement, she asks students to find an equation that satisfies the "if" part of the statement but not the "then" part. After students complete this assignment, Ms. Martinez leads them in a discussion about their findings.]

Fig. 20.4. Classroom vignette of a Type 2 technology-based lesson

In the vignette (fig. 20.4), students begin by examining the characteristics of the graphs of a variety of linear and quadratic functions. Students are asked to impose structure on the set of graphs by considering similarities and differences among them and creating groups and subgroups. Variation is expected in students' methods of grouping, and the ways in which graphs might be reasonably grouped can become a topic for discussion. Next, students are presented with a task (matching equations and graphs) that gives them an opportunity to think about the relationships among these representations. Students use their graphing calculators to determine which equation goes with which graph and to make some conjectures about the relationships among equations and graphs. Finally, students write equations that could be used by others to test their conjectures. As can be seen in the vignette (fig. 20.4), the variety of conjectures produced by this type of lesson is very useful. Teachers can collect students' conjectures (whether true or false) and create a class list. Students can select conjectures from the list and attempt to show why they are false or explain why they must be true. Students' ideas about whether conjectures are true and why can be used for class discussions about the nature of proof. When the students and teacher agree that a conjecture has been proved, it can become a "class theorem" to be used in future proofs.

The second Type 2 example, involving properties of quadrilaterals, is similar in some ways to its Type 1 counterpart. Students use the click-and-drag features of the computer software in conjunction with measuring or other tools to make conjectures about the properties of quadrilaterals. In the Type 2 example, however, students choose a shape to study and a means of studying it. In addition, students are instructed to test their conjectures for other shapes, and the "Why or why not?" question encourages reasoning about why certain properties are exhibited by some shapes and not others. Teachers can use the variety of conjectures produced in ways similar to those described for the graphing calculator lesson—they can create class lists and let students choose conjectures to prove.

Students' mathematical activity of the type advocated by NCTM (2000) permeates Type 2 technology-based lessons. These lessons require that students analyze objects and engage in reasoning as they search for patterns and relationships. Students must make some decisions about *what* things to do (e.g., what objects to construct or measurements to take) and about *how* to do things (e.g., how to organize graphs into groups). These lessons go beyond the identification of conjectures, requiring students to consider why certain relationships seem to exist and whether they would hold for other kinds of objects. These lessons typically require students to communicate their reasoning to one another and to the teacher either orally or in writing. Because students produce a wide range of solutions and solution methods, teachers can engage students in evaluating other students' strategies and mathematical arguments. In short, Type 2 lessons seem to have more potential than Type 1 lessons for engaging students in investigating, reasoning, and developing notions of proof.

Conclusion

"When technological tools are available, students can focus on decision making, reflection, reasoning, and problem solving" (NCTM 2000, p. 24).

The technology-based lessons we described in this article are ones that we have used with secondary school mathematics students in an effort to incorporate appropriate uses of technology into the classroom. Our experiences with students suggest that some kinds of technology-based lessons are more useful than others for encouraging reasoning, conjecturing, and developing mathematical arguments. We used to simplify investigations for our students by selecting one or two objects or relationships for them to study and writing step-by-step directions for them to follow. Now, we deliberately choose larger and more complex sets of objects and require that students take some responsibility for making decisions about which mathematical objects or relationships to examine and how to examine them. In addition to offering students more opportunities to develop their reasoning, problem-solving, and communication skills, these more complex technology-based lessons (which we call Type 2 lessons) are useful in another way. We can collect the conjectures and arguments produced by students and re-present them to the whole class for further examination and evaluation.

In conclusion, we would like to offer readers who are interested in using Type 2 technology-based lessons in their own classrooms a few suggestions related to implementation. We have found that the more comfortable we are with a mathematical topic and associated technology, the easier it is for us to allow students to share in the decision-making process and, at the same time, guide their investigations in mathematically fruitful directions. We have also found that students may not be accustomed to making the decisions required of them by Type 2 lessons. They sometimes say, "Tell me what to do next." We usually reply by asking, "What have you done so far?" and then encourage them to move forward by saying, "Next, you should decide … [e.g., how to categorize the graphs, for which kinds of shapes your conjecture holds, in what ways the parameters are related]." Students often simply need us to reaffirm that the decisions are theirs to make. Finally, if the lesson extends over several days, we typically ask students to spend about ten minutes at the end of each class period writing a short description of what they have done so far and what, specifically, they plan to do next. This helps us keep abreast of students' progress and offer guidance where needed.

When reflecting on the quality of our own technology-based lessons, we find it useful to consider the following questions:

- How much time do students spend trying to understand and follow directions versus trying to identify patterns and find relationships?

- How much time do students spend reasoning about the mathematics versus trying to get the technology to work?

- How much variation is there in the generalizations or conjectures that students make?

- How much time do students spend testing, refining, and attempting to prove their conjectures?

- How much time do students spend waiting for the teacher to help them with the technology?

- How much time does the teacher spend discussing mathematics and methods for investigating mathematics with students versus helping students interpret and follow directions?

Our experiences with technology-based lessons suggest that teachers who frequently use technology in the classroom, as well as teachers who have just begun to use it, can benefit from reflecting on these questions. When we began using technology in the classroom, we reacted to the complexity of the new learning environment by attempting to "keep everyone on the same path." Our own reflections on the first two questions, that is, our frustration with the amount of time our students spent attempting to follow our directions and the amount of time they spent trying to get the technology to work, led us to realize that we needed to move in the opposite direction; that is, we needed to "open up" our lessons and allow for multiple investigation paths. Allowing students to make some decisions about what to do and how to do it helped create richer learning experiences for them and also helped create more manageable classroom environments for us. We have continued to use these questions as a tool for reflection whenever we design and implement technology-based lessons. When designing lessons, we consider whether students will have opportunities to make decisions about what to do and how to do it, whether variation in conjectures is likely, and whether students will be required to test, refine, or prove their conjectures. During implementation, we reflect on all the questions listed above, with particular attention paid to our own interactions with students. In conclusion, we should note that sometimes interpreting directions and figuring out how to use technology can be productive, mathematical activities. Students develop an understanding of the language of mathematics in part by decoding instructions, and solving technology-related problems can help students develop their logical-thinking abilities. Questions such as those listed above are useful as tools for reflection, however, because they help us think about using technology in the classroom in ways that maximize students' opportunities to learn.

REFERENCE

National Council of Teachers of Mathematics (NCTM). *Principles and Standards for School Mathematics.* Reston, Va.: NCTM, 2000.

21

Engaging Students in Authentic Mathematics Activities through Calculators and Small Robots

George C. Reese   Jennifer Dick   James P. Dildine

Kathleen Smith   Mikkel Storaasli   Kenneth J. Travers

Susan Wotal   Dalia Zygas

A UTHENTIC uses of instructional technology can lead to increased motivation and understanding in mathematics. This article describes our experiences with sixth-grade students, high school students, and preservice and inservice teachers as they used small, calculator-based robots. A remarkable discovery made in the course of these activities is that students at a variety of achievement levels are equally engaged. Indeed, it is often difficult to tell, from classroom observation, which students are advanced, which are remedial, or which have learning disabilities. We include information on the tools, how they were used in middle and high school classes, sample programs for the robot activities, and links to video clips of students and teachers engaged in the projects.

A New Way to Look at Authentic Activities

What is an "authentic" task or activity? Means and Olson (1994) discuss authentic tasks as those that "are completed for reasons beyond earning a grade. Students also see the activity as worthwhile in its own right" (p. 17). In our own work in the classroom, we have found that students come to see technology as a powerful tool for working with complex, authentic tasks. We add to this definition that authentic activities are those during which a sophisticated observer cannot tell the difference between the mathematical achievement levels of students by observing the level of interaction with the activity. We have found that, typically, students who are working with robots are intensely focused, regardless of their achievement level in mathematics or

---

Editor's note: The CD accompanying this yearbook contains an electronic version of this article, complete with hyperlinked URLs.

their learning style. Furthermore, it has been our experience that students with limited English language proficiency, learning disabilities, and behavior disorders have all been deeply engaged in the activity. The quality of students' classroom work and the depth of their understanding do, of course, differ, but not their level of engagement. Thus, we see tools like these as providing one way to help *all* students learn. We have seen this occur with a number of handheld technologies, including graphing calculators (Dildine 1999) with distance sensors and other probes, global positioning devices, and small robots. In this article we focus on small robots.

What Are the Robots?

The robots used for our classes are connected to Texas Instruments graphing calculators. We used TI-73 and TI-83 Plus calculators. The robots themselves were designed by Norland Research (Rowland 2003). The robot and calculator combination costs about $200. Since the robots can be used at multiple achievement levels, a set of ten or twenty can be shared and used by several classrooms and schools.

The robot obeys numerical commands sent through the calculator. For example, the command "122, 500" tells the robot to go forward for five seconds. The "1" tells the robot to execute the command for a timed period, the first "2" tells the left wheel to go forward, and the second "2" tells the right wheel to go forward. The "500" is the time, in centiseconds, that the robot will move. Sample programs are available at www.smallrobot.com/testpage.html (see fig. 21.1).

Other types of robots are available, and the technology is popular and evolving. Lego Mindstorms has robots that include light and touch sensors, as well as all the construction activities that go with Lego toys (Lego Group 2000).

```
:For(K,1,6)
:Send({100,200})
:Get(R)
:Send({101,83})
:Get(R)
:End
```

Fig. 21.1. Modifying the program that draws a hexagon. The program appears on the right. Changing the time of the turn changes the angle that makes the hexagon close. The program is available from www.smallrobot.com/testpage.html.

Klutz, Inc., produces a solar-powered car (Klutz 2001), but it is not programmable. These are just two options. A quick search of the Web reveals everything from snake-shaped robots to discussions of the robot rovers on Mars. The Norland robot we use has the following advantages: (1) it uses the Texas Instruments graphing calculators that are common in mathematics classrooms, and (2) it requires only a simple set of commands to begin making the robot work. The capabilities of all these robots will doubtlessly continue to improve.

As the technology improves, we expect the pedagogical advantages to improve as well. We look forward to using robots that record the distance they have traveled and that can display their speed, or robots where the wheel speeds can be adjusted precisely through mathematical functions entered by the user. A current challenge, for example, is to make the robot move in a sine wave. Nonetheless, part of the interest inherent in using these devices is trying to find ways to work around obstacles and limitations. For example, our students have jury-rigged pens to follow along beside or in front of the robot to trace its path. Recent models of the robot now include an arm to hold a pen.

Why Robots?

Reasoning and problem solving are essential components of *Principles and Standards for School Mathematics* (National Council of Teachers of Mathematics [NCTM] 2000). Robots represent a unique opportunity to engage students in designing algorithms that have an immediate, tangible response. Because students (and teachers!) can see the immediate and active results of their work in the motion of the robot, they write programs to make the robot go, and they modify their programs to make it do more and more complex motions. This trial-and-error approach works for students as they learn the commands that make the robot go. But opportunities to use formulas, make measurements, analyze data, and recognize patterns soon emerge.

Learning about the Robots

This section is based on the work of two middle school teachers, Jennifer Dick and Susan Wotal, and two high school teachers, Mikkel Storaasli and Dalia Zygas, who have used the robots in their classes.

The lesson for sixth graders lasts just over a week (Wotal and Dick 2002). Two days are spent becoming familiar with the robots. During the first day, the instructors demonstrate the robot motions and explain the commands to the students as a group. Students observe the robots moving, bumping into an object, backing up, turning, and going forward. They learn the commands and plan what they want the robot to do (fig. 21.2).

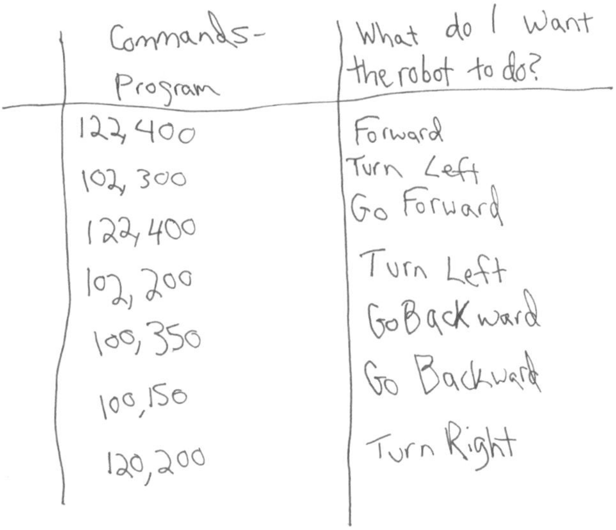

Commands– Program	What do I want the robot to do?
122, 400	Forward
102, 300	Turn Left
122, 400	Go Forward
102, 200	Turn Left
100, 350	Go Backward
100, 150	Go Backward
120, 200	Turn Right

Fig. 21.2. A middle school student's worksheet illustrates one approach to learning the robot's commands. The student lists the commands and what the robot does with each command.

By the second day, they work with partners to program the robots. By the end of the week, students will prepare a visual representation of their final project. In the assessments, students are asked to describe the activity in their own words and to reflect on it. They describe the strategies they used to make the robot move, and they have a chance to state some of the frustrations they had in making the robots move. Some students prepare presentations on the computer or make posters that include pictures of the paths traced.

Figure 21.3 shows the project of one pair of students who made a hexagonal STOP sign. (Oops, STOP signs are typically octagons—another teachable moment with the robots.) To program their robot to travel this path, the students had to determine what angle they wanted the robot to turn and also find the relationship between the length of time the wheels turn and the angle that the robot turns. A student may get the hexagon to close simply by trial and error, but then the teacher can ask, "What would we need to do to the program to make a hexagon with sides twice as long?" The students would then discover that although they needed to increase the time that the robot moved forward, they would not have to change the amount of time that the robot turns. Thus, they have a visible and moving example of the fact that the sum of the interior angles of a polygon is the same regardless of the size of the polygon.

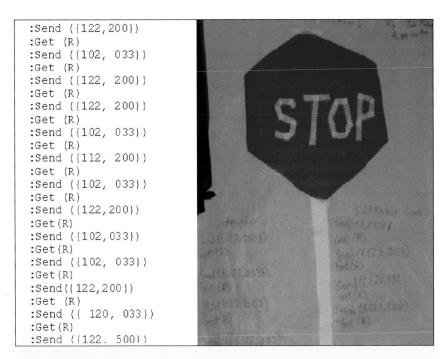

```
:Send ({122,200})
:Get (R}
:Send ({102, 033})
:Get (R}
:Send ({122, 200})
:Get (R}
:Send ({122, 200})
:Get (R}
:Send ({102, 033})
:Get (R}
:Send ({112, 200})
:Get (R}
:Send ({102, 033})
:Get (R}
:Send ({122,200})
:Get(R}
:Send ({102,033})
:Get(R}
:Send ({102, 033})
:Get(R}
:Send({122,200})
:Get (R}
:Send ({ 120, 033})
:Get(R}
:Send ({122. 500})
```

Fig. 21.3. This poster is the final product of a student in a sixth-grade class. The program that traced the stop sign is on the left.

The high school lesson plans (Storaasli and Zygas 2002) were done with students in a mixed mathematics and science classroom. Even though the students are older, the lessons proceed in much the same way as in the middle grades. The first day is spent learning how the robot works. "How do I get it to move forward?" In our experience, it takes about an hour to become familiar enough with the robot to be able to feel confident in programming it. This amount of time is similar for middle school students, high school students, and teachers themselves.

At the high school level, students jump into the mathematics more quickly. Like the middle school students, they work on making a closed polygon and the relationship between the time of the turn and the angle swept out. But they are able to do a more sophisticated analysis. They create a chart that shows the relationship between the time in centiseconds that the wheel is moving and the angle it makes. Is the relationship perfectly linear? If not, what is the line of best fit? Calculating and plotting lines of best fit is done with the same calculators that power the robots. By plotting a regression line, they can see how close to a linear relationship the time of the turn is to the angle swept out. Armed with this discovery, they can write a program that will take input to draw an "*n*-gon" with a user-entered number of sides. Another relationship between the time of travel in a straight line and the distance is also linear. Will those data be more precise or less precise than the angle data? How will you know?

Measurement and pattern recognition run throughout the activity. Students must systematically change the time of the turns and gather enough data to allow them to determine a relationship(s). The data should be collected for a wide range of values to assist in determining the mathematical relationships. Data from a set of angle measurements appear in figure 21.4.

Angle (degrees)	Turn time (centiseconds per turn)	Straight line distance (cm)	Drive time (centiseconds of both wheels going forward)
120	157		
90	118	16	100
72	94	31	200
60	78	46	300
56	67	61	400
45	59	76	500
40	52	90	600
36	47	106	700
32.7	43		

```
LinReg
 y=a+bx
 a=-1.13092105
 b=1.314443157
```

```
LinReg
 y=ax+b
 a=6.697797922
 b=-7.608844939
```

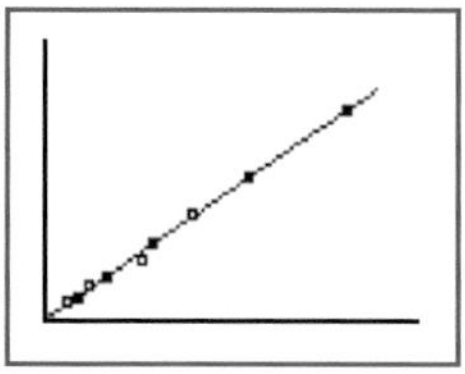

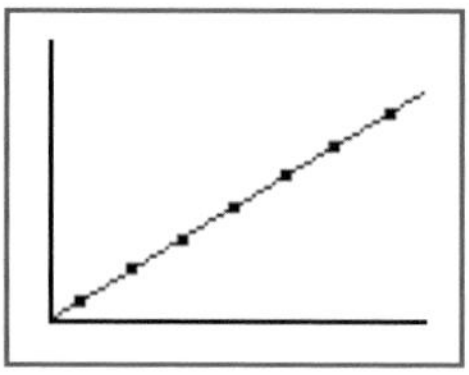

Fig. 21.4. Sample data, scatterplots, and regression calculations. Students use their calculators to create a line of best fit that relates the turning angle of the robot to the amount of time it turns. They also do this for the distance traveled. Details can be seen at www.mste.uiuc.edu/ courses/ci399ATGfa02/folders/mstoraasli/day_3.htm.

Teachers Bring Out the Mathematics

A crucial role of the teacher is to help set up the problems and bring out the mathematics embedded. The activities done by Wotal and Dick (2002) and by Storaasli and Zygas (2002) require students to measure and estimate, to collect and analyze data, formulate their reasoning, and communicate their results to others. They must accurately determine the linear distance-time relationship to be able to predict the location of the robot at a particular time. With students and teachers working imaginatively, even more connections are available. For example, if you were to attach a distance sensor to a robot, what would the time vs. distance graph look like as the robot moves toward a wall? (*Answer*: It is a straight line.) Can you program the robot to draw one of the letters of your name? Can you make it do a three-point turn? Can you make it parallel park? If a marker following the left wheel makes one shape (e.g., a circle, a certain polygon, or a flower shape), predict the shape made if the marker is placed on the other wheel. Have your robot make turns of different angles. Make a table that shows the relationship between the amount of time for the turn and the angle of the turn. Use the data to make a graph that shows "Time of turn" vs. "Angle of turn."

Professional Development with the Help of Robots

We have found that teachers also become very interested with using the robots, though adults tend to be more intimidated by esoteric commands to calculators than the increasingly technologically savvy students of today. Robots present a remarkable opportunity for teachers at a variety of grades to share in an inquiry-based activity, such as one of the following:

- On a piece of 22″ × 28″ poster board, draw a path. Now program your robot to follow that path.
- Take the program for drawing a hexagon and modify it to make an octagon.
- If your robot is running a program to wander around an empty 30′ × 20′ room for twelve hours, estimate how far it will travel (assuming the batteries do not run down).

Conclusion

Working with robots is more than just fun for the students and teachers. Like the Logo turtle or the Java turtle in an NCTM Illuminations activity (NCTM and MarcoPolo 2003–04), a robot helps students increase their conceptual knowledge of geometric properties. But the robot adds the additional element of being a physical object that stops, starts, and follows commands. Jury-rigging the robot with markers adds the additional motivation that comes with invention. We have found that students generally want to make it do more.

Principles and Standards for School Mathematics (NCTM 2000) recommends that students be able to "solve problems that arise in mathematics and other contexts" (p. 53). Robots can be a part of a mathematics curriculum that emphasizes problem solving. These problems arise as students try to create algorithms to perform tasks such as drawing polygons. In following mazes, creating polygons, or predicting outcomes of different algorithms, for example, they are forced, as recommended in *Principles and Standards,* to find the "appropriate techniques, tools, and formulas" (p. 242) to make the robot do the required task.

As an added benefit, when students see robot vacuum cleaners and robots that work in factories, their experience with these small, calculator-based robots will give them a reference point for deeper understanding about the processes that control these machines.

Examples of Robot Programs

Note: The speed of the wheels varies from robot to robot. Gather data to determine the rate at which your robot moves and turns.

Equilateral triangle—Adapt this program to move a robot in the path of an equilateral triangle.

```
:For(K,1,3)
:Send({122,300})
:Get(R)
:Send({112,157})
:Get(R)
:End
```

A 90-60-30 triangle—The table below shows the data for the distance traveled and the angles of turn for a particular robot. From these data, verify that the program below will approximate a 90°- 60°-30° right triangle.

Turn command = 112	Time of turn in centiseconds	Angle measure at left wheel (exterior angle)	Time of travel (centiseconds)	Distance traveled (cm)
	50	55	100	16.5
	100	93	200	32
	150	130	250	39
	200	165	300	46.5

```
:Disp "Program Make 90 60 30"
:Send({122,165})
:Get(R)
:Send({112,100})
:Get(R)
:Send({122,100})
:Get(R)
:Send({112,135})
:Get(R)
:Send({122,200})
:Get(R)
```

Explain why the following program makes a triangle similar to the one above.

```
:Disp "Make Larger 90 60 30"
:Send({122,2*165})
:Get(R)
:Send({112,100})
:Get(R)
:Send({122,2*100})
:Get(R)
:Send({112,135})
:Get(R)
:Send({122,2*200})
:Get(R)
```

Video Clips

Videos of students and teachers working with the robots are available online through the following links and from links on the CD accompanying this yearbook.

- www.mste.uiuc.edu/reese/nctm/triangles.avi (6.8 MB) shows students creating different triangles with the robots.
- www.mste.uiuc.edu/reese/nctm/maskingTapeMaze.avi (5.6 MB) shows students attempting to program a robot to follow a given path.
- www.mste.uiuc.edu/reese/nctm/SummerMathRobots.avi (5.6 MB) shows a variety of activities that students do during summer workshops with the robots.
- www.mste.uiuc.edu/reese/nctm/rightTriangle.avi (2.9 MB) shows a robot drawing a right triangle on a sheet of marker board.

- www.mste.uiuc.edu/reese/nctm/TooShort.avi (2.3 MB) and www.mste.uiuc.edu/reese/nctm/JustRight.avi (6.8 MB) contain short videos of teachers working with robots trying to get the robot to follow a loop pattern.

REFERENCES

Dildine, James P. "Technology-Intensive Instruction with High-Performing and Low-Performing Middle School Mathematics Students." Master's thesis, University of Illinois, 1999. Available at www.mste.uiuc.edu/dildine/thesis/index.htm.

Klutz, Inc. *The Solar Car Book: A Complete Build-It-Yourself Solar Car Kit Including All the Parts, Instructions and Pain-Free Science.* Palo Alto, Calif.: Klutz, Inc., 2001.

Lego Group. *Lego Mindstorms.* Billund, Denmark: Lego Group, 2000. Available at www.legomindstorms.com/.

Means, Barbara, and Kerry Olson. "The Link between Technology and Authentic Learning." *Educational Leadership* 51, no. 7 (1994): 15–18.

National Council of Teachers of Mathematics (NCTM). *Principles and Standards for School Mathematics.* Reston, Va.: NCTM, 2000.

NCTM Illuminations, and MarcoPolo. "Developing Geometry Concepts Using Computer Programming Environments." Illuminations lesson activity. 2003 [cited September 28, 2003]. Available at illuminations.nctm.org/index_d.aspx?id=396.

Rowland, R. Graphing Calculator Kit for Small Robots. Las Vegas, Nev.: Norland Research, 2003. Available at www.smallrobot.com.

Storaasli, Mikkel, and Dalia Zygas. "Robot Cars, Data Collection and Analysis and the TI-83." [World Wide Web]. Office for Mathematics, Science, and Technology Education (MSTE), 2002 [cited March 12, 2003]. Available at www.mste.uiuc.edu/courses/ci399ATGfa02/folders/mstoraasli/robot_lesson.htm.

Wotal, Sue, and Jennifer Dick. "Using Your Robot." [World Wide Web]. Office for Mathematics, Science, and Technology Education (MSTE) 2002 [cited September 28, 2003]. Available at www.mste.uiuc.edu/summermath/2004/activities/robots/appleworks/.

22

Using GIS to Transform the Mathematical Landscape

Bob Coulter
Joseph J. Kerski

FOR the past forty years, teachers have been seeking ways to capitalize on the power of computer technology to improve education. More generally, efforts to use technology to enhance learning date back nearly a century (Cuban 1986). Although the specifics of each form of technology are different, the belief remains that the right tool, correctly employed, will lead to vastly improved learning opportunities. As a result, many schools and districts have invested heavily in computers and technology-focused professional development for teachers.

What is often lacking in this rush to bring technology to education is a critical perspective that asks about purposes and trade-offs: Why this technology? Toward what ends is it expected to work? What will be displaced by its presence? Without this questioning, each innovation is brought in with great fanfare, only to be replaced by the "next big thing," which inevitably is only a year or two down the road. It wouldn't take too much mental energy to imagine a museum of technology innovations that have, by and large, failed to live up to their claimed potential to improve education on a broad scale, including programming, multimedia software, and data exchanges among schools. Nearby might be a museum of still common, but largely underused technologies, including the use of the Internet for research and spreadsheets. These latter examples, although still important players in the field, have yet to realize the widespread, revolutionary impact that their advocates have promoted so fervently.

Within this daunting context, why should a comparatively new technology—a geographic information system, or GIS—be considered? In an environment already filled with technology applications, why add another one? To answer this, we invite you to look to the near future, inside a radically transformed mathematics learning environment—one that addresses several well-

---

Editor's note: The CD accompanying this yearbook contains hyperlinked URLs for the Web sites that are relevant to this article as well as a GIS tutorial.

known educational problems and one for which the *need* for GIS will become evident. Just as you don't go to the hardware store to pick up a new hammer and look for things to pound, you shouldn't pick up GIS (or any other technology) and look to see what problems it can solve for you. Inevitably, such technology grafting will create as many or more problems than it solves. Ultimately, technology is a tool or a resource, suitable for specific situations where the work either couldn't be done without the tool or couldn't be done as well. In a classroom that embraces the *Principles and Standards for School Mathematics* (NCTM 2000), there will be many situations in which GIS will be needed. With the need apparent, integration will then have a much greater likelihood of sustaining meaningful change.

What Is a Geographic Information System?

GIS software, available in either free basic versions or in fully featured versions with a schoolwide site license for as little as a few hundred dollars, is offered by several developers and for several computer platforms.[1] The software allows the user to integrate data about the location of something and its attributes. Your hometown and other cities can be located using a coordinate system such as latitude and longitude and plotted on a map. Local climate data for these cities—such as average monthly temperature and precipitation—can also be plotted with symbols on the map from a data table, allowing an analysis of patterns and trends among variables. For example, how does latitude (or elevation) affect temperature? GIS allows students to pursue these mathematically based, real-world questions.[2]

The power of GIS exists in its unique ability to integrate geographic data, or data about location, with specific attributes. Continuing the climate example, no one should be surprised that the average January temperature in Yuma, Arizona (56.5 degrees), is warmer than the temperature for the same period in International Falls, Minnesota (1.0 degree). However, careful analysis reveals considerable regional variation in the data, due to a range of factors. Although most people would expect that Minnesota is colder than Arizona, fewer people know that on average, it is colder in parts of Arizona than in Denver in January (fig. 22.1).

---

1. Major developers include ESRI (www.esri.com/k-12), Intergraph (www.ingr.com), Clark University (www.clarklabs.org/), and MapInfo (www.mapinfo.com). General information about GIS in education and links to additional resources can be obtained at the U.S. Geological Survey site (rockyweb.cr.usgs.gov/public/outreach).

2. A sample GIS tutorial exploring climate patterns is included on the CD-ROM accompanying this book.

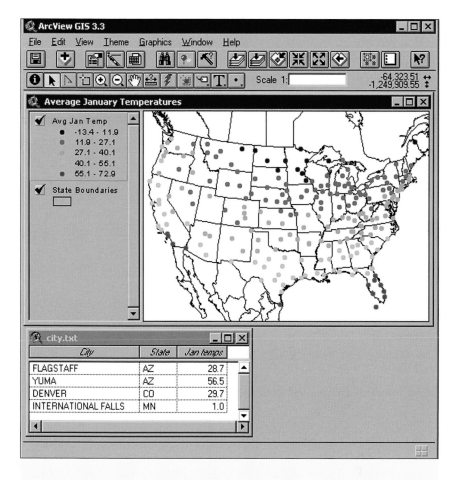

Fig. 22.1. A GIS-produced map and table of average January temperatures across the continental United States

There is a world of difference between the student who makes a generic claim that it is warmer in the south than in the north and the one who is able to quantify how much warmer it is, interpret the data according to regional patterns, and notice variations such as the relative "warmth" in Denver. The latter student is becoming a mathematically literate citizen. Students using GIS can build toward this literacy as they gain practice in analyzing both the kind of data captured in variables—such as temperature or elevation—and relationships among these variables on maps and in tables of data. As will be demonstrated, using GIS does not simply involve applications of concepts already learned. Under the leadership of a capable teacher, the mapping and data analysis tools that make up a GIS can be used to develop, refine, and extend the skills and dispositions that constitute mathematical literacy.

A Day at Feldman Middle School[3]

After a concerted effort to align its curriculum and teaching practices to the visions found in the NCTM *Principles and Standards for School Mathematics* and standards documents from other disciplines, Feldman has found that its students' learning experiences are dramatically different from what they were in the past. This change did not happen overnight, but each year the school has moved closer to the ideal of an academically rigorous, carefully sequenced, and meaningful curriculum. The faculty has also developed a consensus on a common set of instructional technology resources to which students will have access as needed to support their investigations. This mutual understanding among the faculty of pedagogy, curriculum, and resources has advanced the level of students' learning experiences considerably.

The morning begins in social studies class, where eighth graders are now studying population shifts in U.S. history. The students are in the computer lab today during a ninety-minute block period, working with a map of county boundaries and a table of data showing the population in each county in the United States for each decennial census from 1900 to 2000. Each team of students has chosen a question from a list the class developed earlier in the week. One group has chosen to investigate which counties have gained the most population; another is exploring to see if any counties actually lost population in the twentieth century.

The group investigating the counties with the greatest population growth constructed a new column in their table and created a formula that enabled the GIS to calculate the difference in population from 1900 to 2000. They then selected the column of data containing the population growth and sorted the list from greatest increase to the smallest, which were actually decreases in population. This exercise gave the students practice in generating meaningful equations and provided an authentic use of negative numbers. Thinking they were done, the students proudly created a graph and a map showing the top ten counties in added population. Ms. Knosp, their teacher, knew that the group was capable of much more substantive inquiry, so she gave them two challenges:

1. Identify other ways to measure population growth.

---

3. Although this vignette is fictional, it is based on the authors' work over the past ten years in a wide range of classrooms integrating GIS technology to support students' learning.

2. Identify a pattern in the counties that were experiencing the greatest growth.

Ms. Knosp then checked in with the group that was looking to see if any counties had actually lost population. Like the first group, these students used their GIS software to conduct a simple calculation to determine that there were in fact more than 600 counties that had lost population since 1900. Ms. Knosp knew that this group wasn't as prepared for the higher-level questions that the first group was wrestling with, so she challenged them to find out just how much the overall population in the United States had increased since 1900, and to see if there were any geographic patterns in the counties that had lost population.

The first group was a bit puzzled by what other methods one could use to measure population growth, so when Ms. Knosp circled back to them later in the class period, she gave them a brief lesson on how they could use their knowledge of ratios. With their lesson framed in this way, the students came to a different set of conclusions about which counties had the greatest increase in population. Their initial set of results—based simply on measuring the total number of people added over time—proved to be quite different from the results they obtained by calculating ratios between the population in 2000 and what the population was in 1900. Many counties didn't add as many individual people in the twentieth century as the large cities but still saw their populations multiply many times over.

Both groups concluded their investigations by looking for patterns in the data they studied. This helped to focus attention on major trends in twentieth-century U.S. history, including the growth of industry in the Northeast early in the century and the growth in Florida and Arizona later in the century. Students investigating population declines also saw a strong pattern in their data, with most of the counties that lost population being in agricultural areas of the country (fig. 22.2).

Moving to science, the students resumed an investigation of seismic activity. They had read about earthquakes in their textbook and worked with some models to gain a basic understanding of important concepts. Their science teacher, Mr. Kilburn, believed that with this base of experience, they were now ready to extend their understanding with focused data analysis. The class began with a challenge to predict where recorded earthquakes of magnitude 2 or less would be found. After a few minutes of discussion, each group reported their prediction, along with an explanation of why they believed this would be the pattern found.

As they engaged in this process, students were required to draw on their background knowledge about Richter scale measurements (where lower numbers represent minor seismic disturbances) and general patterns in global

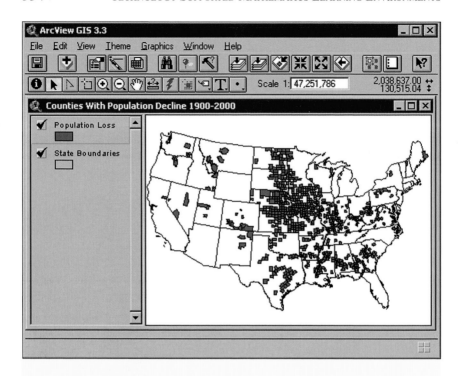

Fig. 22.2. Map resulting from asking the GIS which counties declined in population during the twentieth century

earthquake activity. A variety of predictions were made, with reference to patterns that the students observed in maps showing all the recorded earthquakes from last year. After some discussion about the different predictions, the teacher modeled how to use a filtering process with the GIS software to change the map projected on the screen from showing more than 46,000 recorded seismic events down to 25,534 recorded events of magnitude 2 or less. The high number of events surprised the students, who had read that the earth was always experiencing seismic events, but they were now confronted with data showing that more than 25,000 earthquakes were listed that were too minor to be felt, clustered on the Pacific coast of the United States, in western Europe, and in New Zealand (fig. 22.3).

When confronted with these data, students were asked to develop a hypothesis about the resulting pattern. Students struggled with this for a while before the teacher reiterated the question but emphasized the word recorded. At this point, several students suggested that there are most likely other small events around the world but that they may not be recorded in the

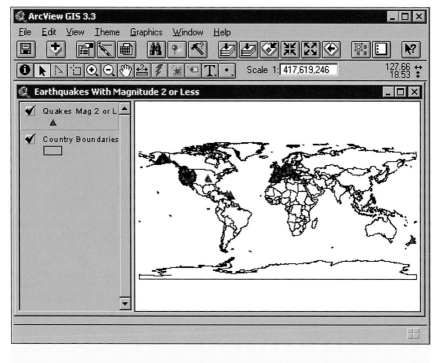

Fig. 22.3. World map showing small-magnitude earthquakes

data set that the students were using. This led to a discussion of the location of seismic monitoring stations that were able to sense minor disturbances. Pedagogically, this interchange led to students' developing a better sense of how and why it is important to be critical of the data being analyzed, always probing for what is not shown.

After lunch, the students moved to their mathematics class, which was studying measurements. Mr. Strauss, the mathematics teacher, coordinated with the science teacher in the content focus and was using GIS and the same data set on earthquakes for examples that furnished a "real world" connection to the work the students were doing. The warm-up challenge for the day was for students to apply what they knew about the San Andreas Fault—including that the two plates at the fault were moving past each other at about two inches a year—to determine when Grants Pass, Oregon, would be adjacent to the current Mexican border. Using the measurement tools in the GIS software, students measured how far Grants Pass is from the border and worked to determine their answer to how long it would take for the two locations to reach each other. As a refresher, this exercise was productive, since the students' skills in the use of measurements and conversions, fractions, and arithmetic computation were strengthened.

After a short work period, students paired off to compare answers and resolve differences. Even after calculation errors were corrected, the students were surprised to find that there was such a wide range of results. Some students argued that a point along the current Mexican border would reach Grants Pass in just over 24 million years, whereas others obtained a result of almost 27 million years. Although the specific time frame the students arrived at was not crucial, Mr. Strauss used the opportunity to lead a very productive discussion among the students about the source of the different answers. Students soon came to realize that the calculations and conversions they were doing were the same, and the difference in the answers arose as a result of the different distances from Grants Pass to the border. Some used a straight-line measurement (about 765 miles), whereas others followed the course of the fault as it was represented on the map (about 840 miles).

Building on this observation, Mr. Strauss made a transition into the algebraic concepts students had been working with previously. He challenged the students to create an algebraic expression that would allow any distance measurement entered into the equation to return a value representing the number of years it would take for two points along the fault to meet. After modeling the type of work that would be needed using simpler cases, he allowed the students to work individually to develop their equation and then check their answer with a neighbor. Throughout this process, Mr. Strauss was able to help his students build an understanding of the power of mathematics to represent a generalized relationship from the very specific problem he posed at the beginning of the class period.

***

Each of these classes is characterized by strong, age-appropriate content and a balance between teacher- and student-driven learning; and, most relevant for this discussion, each involves a significant degree of spatial analysis enhanced by the use of GIS. This enhancement added value to the lessons beyond what could be done with a spreadsheet. Their analysis of the geographic patterns in the data enabled students to connect their analysis of population to their understanding of twentieth-century U.S. history. Spatial analysis also helped to interpret earthquake patterns and to solve a challenging measurement problem by visualizing the situation and collecting relevant data not easily found elsewhere, such as the distance between two points along the San Andreas Fault.

Using GIS to Meet Curriculum Standards

In our work, we have found that thoughtful use of GIS makes significant—and in many instances, unique—contributions to the mathematics learning environment. The spatial and quantitative tools in a GIS enable broad and deep learning in the ten strands described in the *Principles and Standards for School Mathematics* (NCTM 2000). The examples cited here are only illustrative; many other applications are also possible.

Number and operations concepts are at the core of many investigations. Students' capacity for effective *use* of these concepts will be developed considerably through their application in a range of contexts. At Feldman Middle School, students experienced many authentic uses of number and operations, including arithmetic calculations; uses of fractions, ratios, and percentages; scales such as Richter; and linear measurements. Each of these promoted an extension of basic skills in a context much more meaningful than the use of review sheets or games.

Algebra skills are enhanced through the use of the data analysis tools in a GIS as students investigate and describe change, such as in climate or population data. For example, one could plot the national growth in population over time, or the population growth in a specific county or state. Dramatic changes in population are represented by a steep slope, whereas gradual changes have a more moderate slope. In the vignette described at Feldman, algebra enabled students to describe a general relationship between two variables they were studying (distance along the plate boundary and time). Students are much more likely to see the value of algebra when they use it in authentic projects, avoiding the familiar "When are we ever going to use this?" refrain.

Geometry skills are developed as students map and analyze coordinate points representing certain specific locations, such as earthquake epicenters or weather stations. Similarly, GIS can be used to explore patterns in how many tornados strike in their state each month compared with other regions, and seeing seasonal patterns in tornado strikes across the country. These processes are fundamental to developing sound mathematical thinking. As Wigglesworth (2000) noted in his research, spatial analysis—an ability "connected to mental rotation, pattern recognition, and the interpretation of three-dimensional space" —is enhanced as students gain experience with mapped data. Numbers take on new meaning when they are not simply an abstraction but a representation of a particular phenomenon, located at a specific place relative to other, comparable measurements. A comparison of these measurements relative to their positions is at the heart of spatial thinking.

Measurement is nearly always an integral part of an investigation. Quite often, there is a geographic or spatial pattern to the measurement process, as seen previously in the recording of minor earthquakes or in the historical issue of enumerating slave populations. More generally, students need to understand the data being used, including what the data represent and how the data were collected. For example, within the context of the social studies class described in the vignette, students should know that census data are collected using a standard protocol. Building on this information, they should also know that some people are less likely to be counted, and they should have some understanding of what the significance of this undercounting may be.

Data analysis and probability skills permeate GIS-enhanced investigations. With a strong understanding of what the data represent, students can engage in meaningful data analysis, ranging from describing central tendency to deciding how to handle outliers. Statistics teachers will find many useful tools in a GIS, such as the ability to classify data into groups by standard deviation, equal intervals, or quartiles. Looking beyond specific tools or procedures, students (and all citizens) need to be critical consumers of what the numbers in a data set actually represent and how they are being reported and analyzed. Throughout, as questions are formulated and appropriate methods of analysis are chosen, data literacy skills are bolstered. Also, many data-rich investigations using GIS enable mathematical modeling opportunities, such as projections of the economic and ecological benefits of preserving trees in a neighborhood (Hagevik 2003).

Problem solving is supported through students' engagement with authentic phenomena as they try to make sense of what they are experiencing. An analysis of earthquake data offers students opportunities to investigate their local risk and to see how mathematics is used to help scientists monitor geologic processes. Likewise, investigations of population data invite the consideration of issues such as the best method for the reapportionment of representatives every ten years based on the census data (Monmonier 2001). In order to maintain a focus on problem solving, effective tools such as a GIS are needed; otherwise, the management of the data becomes a larger task than the problem to be solved.

Reasoning and proof capacities are enhanced as students develop explanations and justifications for patterns they observe. In the vignettes described in this article, students needed not only to identify patterns in the population changes they saw but also to offer tentative explanations that were guided or characterized by the data and by other background knowledge. In the study of minor earthquakes, students had to develop a hypothesis based on their understanding of earthquakes, and when presented with the actual data for small earthquakes, many had to revise their thinking to better fit the patterns they observed. Through spatial analysis, a more carefully reasoned perspective is developed that is characterized by the data. This perspective in turn can be used to establish proof for the conjectures that students develop.

Communication skills support mathematics as students develop and express arguments and evaluate the arguments offered by others. In order to interpret data, a range of communication skills is required. Background reading contributes to a deeper understanding of what the data represent, and students have to be able to "read" data displays. As this interpretive process takes place, students articulate their understanding of the data and listen as they evaluate the arguments offered by peers. Often, a revision and reformulation of ideas will be required. Throughout, mathematically based thinking, speaking, and listening occur.

Connections among mathematical ideas are fostered as students solve problems by integrating numbers, graphs, tables, and maps. Instead of being used only in isolated units ("This week we're doing graphs …"), an array of tools and concepts are brought to bear on a problem, supporting an integrated view of mathematics. When specific skills are taught, connections to authentic contexts support, retention and extension. In a class focusing on measures of central tendency, climate data furnish a rich context as longer patterns are contrasted with "point in time" measurements. Most students have a sense of unusually warm or cold weather; having a "normal" measurement to compare this to offers a mathematical way for students to process their experience.

Numeric, graphic, and spatial *representations* of data each have strengths and limitations. Imagine a student during an unusually cold week. The student could view a table, graph, or map to help in understanding her experience. The table would give a concise summary of the data, and the graph would show the phenomena in a more visual manner. A map could be used to view regional patterns and to determine if warmer weather was likely to be coming. As students become critical consumers of different forms of representation, they are better able to clarify their thinking and may be better able to transfer these skills to other situations (Monk 2003).

Looking at the Research: How Does GIS Support Learning?

GIS-enhanced inquiries can address a number of issues raised in *Principles and Standards for School Mathematics*. Since applications of GIS in education are relatively new, a comparatively small amount of research has been done concerning its use (e.g., Kerski 2003), though a number of case studies (e.g., Alibrandi 2003; Audet and Ludwig 2000) and dissertations (e.g., Weller 1993; Wigglesworth 2000; Olsen 2000; and Hagevik 2003) document specific educational benefits, particularly as they relate to students' ability to work with data.

Adding to the research data specific to GIS applications in education, a great deal of support has been offered by generic research on learning (National Research Council 2000; Jackson and Davis 2000) and on achievement in mathematics (National Center for Education Statistics 2000). This research offers strong support for the development of cognitively complex environments that are rich in inquiry and grounded in authentic contexts.

The National Research Council's (2000) book *How People Learn*—seen as a standard reference in the field—notes that "a new theory of learning is coming into focus that leads to very different approaches to the design of curriculum, teaching, and assessment than those often found in schools today" (p. 3). Major findings of that project, both in general learning theory and in applications specific to mathematics education, support the approaches described here. One of the fundamental considerations undertaken by the National Research Council panel was the difference in ways experts and novices approached problems. Much of the problem solving that goes on in classrooms today is closer to the novice end of the spectrum. Given the age of the students, this is not totally unexpected, but it raises the question of what skills will be required to enable students to develop their mathematical expertise. The National Research Council described six major behaviors exhibited by experts, each of which is relevant to the design of mathematically rich environments (p. 31):

1. Experts notice features and meaningful patterns in information that are not noticed by novices.

2. Experts have acquired a great deal of content knowledge that is organized in ways that reflect a deep understanding of their subject matter.

3. Experts' knowledge cannot be reduced to sets of isolated facts or propositions but instead reflects contexts of applicability; that is, the knowledge is "conditionalized" on a set of circumstances.

4. Experts are able to flexibly retrieve important aspects of their knowledge with little attentional [*sic*] effort.

5. Though experts know their disciplines thoroughly, this does not guarantee that they are able to teach others.

6. Experts have varying levels of flexibility in their approach to new situations.

The need for ability in pattern recognition and effective use of prior knowledge is well known, though often this skill is less than optimally achieved. In the many classes we have worked with using GIS, students were able to extract patterns from the data and analyze these patterns using a range of mathematical concepts. Doing this well poses challenges, especially when dealing with "real world" data filled with measurement errors and outliers. With appropriate teacher leadership and the right tools, however, students can make sense of patterns in data. When the data have a spatial component, GIS is usually an essential tool in organizing and viewing the data. Without the spatial display, pattern recognition becomes more difficult, if not impossible.

Careful structuring by the teacher coupled with the analytic capacity of GIS software lends effective support for students' efforts to use their mathematical knowledge. The students in the vignette working with census data were able to set up calculations showing additive population changes and then readily change their calculations to investigate proportional changes before comparing results in tables and maps. The GIS allowed the students to focus on the more significant issue of which is a better measurement for their purpose. Without it, the tedium of rote calculations may have discouraged the students from starting. In the absence of these higher-level experiences, students will not have adequate opportunities to move their thinking along the continuum from mathematical novice to expert.

The last two points raised by the National Research Council also bear consideration, since practically everyone has experienced a class taught by a person who was very knowledgeable in his or her field but who was unable to teach effectively. The opportunities afforded students to describe what patterns they see, listen critically to the perspectives offered by their peers and teachers, and reformulate their thinking all support the pedagogic flexibility required to teach others—an essential skill in a knowledge-based economy (Hargreaves 2003). Also, this reflective thinking encourages flexibility both in the approaches taken for a given problem and in the ways in which one approaches other problems.

Finally, note the similarities in pedagogic approaches in the vignette. Although this exercise condensed students' overall experiences for purposes of illustration, the necessity of repeated experiences in similar but not identical contexts allows students to broaden the ways in which they are able to approach new situations effectively. "Transfer," or how one takes existing skills and knowledge and applies them to different situations, is enhanced when the instruction is active and done in multiple contexts. This transfer must be explicitly planned for and built into the curriculum, with due consideration paid to the tasks students engage in: "With multiple contexts, students are more likely to abstract the relevant features of concepts and develop a more flexible representation of knowledge" (National Research Council 2000, p. 78). As a student moves from novice to expert status, he or she increases the ability to transfer knowledge and understanding to new situations, leading to the development of a deeper mathematical worldview that is at the heart of *Principles and Standards for School Mathematics.*

Summing Up: Why GIS?

Perhaps the most concise summary of why GIS should be included in the mathematics curriculum is found in the consideration of the Technology Principle near the beginning of the *Principles and Standards* (p. 24):

> Electronic technologies ... are essential tools for teaching, learning, and doing mathematics. They furnish visual images of mathematical ideas, they facilitate organizing and analyzing data, and they compute efficiently and accurately. They can support investigation by students in every area of mathematics, including geometry, statistics, algebra, measurement, and number. When technological tools are available, students can focus on decision making, reflection, reasoning, and problem solving.

From our experience with thousands of teachers in a wide range of schools, we have found that the use of GIS as one of several instructional resources enables teachers to create and sustain a learning environment that would be much more difficult—if not impossible—to sustain without GIS. First, the unique integration of both quantitative and spatial data supports the development of a wide range of mathematical capacities, including the understanding of content, processes, and connections to other disciplines. Second, GIS also presents opportunities to enhance the science and social studies curriculum. This is not simple encroachment by the mathematics department on others' domains. In each instance, the infusion of real-world data into the curriculum strengthens the learning potential in those classes and helps teachers move closer to realizing the academic standards in their respective disciplines. Strong mathematics contributes directly to improved science and social studies classes.

Third, the ability afforded by a GIS to bring the real-world data and issues into the mathematics classroom better enables teachers to realize the full depth and breadth of the NCTM's *Principles and Standards for School Mathematics.* Appropriate integration of GIS into the curriculum supports learning in all ten strands identified in the *Principles and Standards.* Because of its ability to use multiple forms of representation and to integrate authentic data, GIS benefits students' understanding of measurement, data analysis, connections, and representations when these investigations are augmented with GIS. A spreadsheet can help a student to manage data, but it cannot enable students to measure on screen (such as was done with the fault line), it cannot help students analyze spatial patterns (imagine looking for the geographic patterns in a table listing 25,000 minor earthquakes), and it cannot help students to generate the range of maps and three-dimensional representations that a GIS can, all of which ultimately supports deeper inquiry.

This is not to claim that GIS is the "wonder tool" of education: We as a profession have gone down that road too often. Instead, consider the tool analogy further. For situations in which students work with data that have a spatial or geographic component, no tool offers the range of capacities that a GIS can. For other problems and situations, different tools may be more appropriate. Just as a hammer or screwdriver offers particular strengths, no tool is best in all situations. As educators, we need to make judicious choices that enable our students to reach their potential. As the *Principles and Standards* notes, "Because of technology … the boundaries of the mathematical landscape are being transformed" (p. 27). We invite you to explore the power of GIS with your students as you transform your classroom landscape.

REFERENCES

Alibrandi, Marsha. *GIS in the Classroom.* Portsmouth, N.H.: Heinemann, 2003.

Audet, Richard, and Gail Ludwig, eds. *GIS in Schools.* Redlands, Calif.: Environmental Systems Research Institute, 2000.

Cuban, Larry. *Teachers and Machines: The Classroom Use of Technology since 1920.* New York: Teachers College Press, 1986.

Hagevik, Rita. "The Effects of Online Science Instruction Using Geographic Information Systems to Foster Inquiry Learning of Teachers and Middle School Science Students." Ph.D. diss., North Carolina State University, 2003.

Hargreaves, Andy. *Teaching in the Knowledge Society: Education in the Age of Insecurity.* New York: Teachers College Press, 2003.

Jackson, Anthony W., and Gayle Andrews Davis. *Turning Points 2000: Educating Adolescents in the 21st Century.* New York: Teachers College Press, 2000.

Kerski, Joseph J. "The Implementation and Effectiveness of GIS in Secondary Education." *Journal of Geography* 102 (May-June 2003): 128–37.

Monk, Stephen. "Representation in School Mathematics: Learning to Graph and Graphing to Learn." In *A Research Companion to "Principles and Standards for School Mathematics,"* edited by Jeremy Kilpatrick, W. Gary Martin, and Deborah Schifter, pp. 250–62. Reston, Va.: National Council of Teachers of Mathematics, 2003.

Monmonier, Mark. *Bushmanders and Bullwinkles: How Politicians Manipulate Electronic Maps and Census Data to Win Elections.* Chicago: University of Chicago Press, 2001.

National Center for Education Statistics. *Mathematics and Science in the Eighth Grade: Findings from the Third International Mathematics and Science Study.* Washington, D.C.: U.S. Department of Education, Office of Educational Research and Improvement, 2000.

National Council of Teachers of Mathematics (NCTM). *Principles and Standards for School Mathematics.* Reston, Va.: NCTM, 2000.

National Research Council. *How People Learn: Brain, Mind, Experience, and School.* Expanded ed. Washington, D.C.: National Academy Press, 2000.

Olsen, Timothy Paul. "Situated Learning and Spatial Informational Analysis for Environmental Problems." Ph.D. diss., University of Wisconsin—Madison, 2000.

Weller, Kay Ellen. "The Appropriateness of GIS Instruction in Grade Six for Teaching Kansas Water Resources." Ph.D. diss., Kansas State University, 1993.

Wigglesworth, John C. "Spatial Problem-Solving Strategies of Middle School Students: Wayfinding with Geographic Information Systems." Ed.D. diss., Boston University, 2000.

23

Technology in Mathematics Education: Tapping into Visions of the Future

M. Kathleen Heid

MORE than a quarter of a century ago, the first microcomputers for use in schools were being marketed; almost twenty years ago, the first graphing calculator hit the market; and fifteen years ago, we saw the first World Wide Web activity. What will technology look like in mathematics classrooms ten years from now? How will technologies be used to affect school mathematics content and teaching? What are the principal changes in mathematics teaching and learning that we might expect over the next decade? The goal of this article is to look forward to what technology-supported mathematics learning environments might look like in the future. To find answers to this question, I interviewed each of the following individuals, all of whom are noted for leading-edge thinking on the uses of technology, most of them in the use of technology in the teaching and learning of mathematics:

> Daniel Chazan, mathematics education faculty, University of
> Maryland
> J. Douglas Child, mathematics faculty, Rollins College (Winter
> Park, Fla.)
> Jonathan Choate, mathematics faculty, Groton School (Groton,
> Mass.)

Jere Confrey, mathematics education faculty, Washington University (St. Louis, Mo.)

Helen Doerr, mathematics and mathematics education faculty, Syracuse University (Syracuse, N.Y.)

James T. Fey, mathematics and curriculum and instruction faculty, University of Maryland

William Finzer, software developer (Fathom), KCP Technologies (Emeryville, Calif.)

Eric Hart, mathematics and mathematics education faculty, Maharishi University of Management (Fairfield, Iowa); past director, NCTM Illuminations project

Nick Jackiw, software developer (Geometer's Sketchpad), KCP Technologies (Emeryville, Calif.)

James Kaput, mathematics faculty, University of Massachusetts Dartmouth

John Kenelly, emeritus mathematical sciences faculty, Clemson University (Clemson, S.C.)

Kenneth Koedinger, faculty, Human-Computer Interaction Institute, Carnegie Mellon University (Pittsburgh, Pa.)

Janet May, program developer, World Campus, The Pennsylvania State University

Robert McCollum, mathematics faculty, Glenbrook South High School (Glenview, Ill.)

Ricardo Nemirovsky, project director, Calculus@Museum–TERC (Cambridge, Mass.)

Kyle Peck, instructional systems faculty, The Pennsylvania State University; founder, Centre Learning Community Charter School (State College, Pa.)

Steve Rasmussen, Key Curriculum Press (Emeryville, Calif.)

Peter Rubba, professor and Director of Academic Programs, World Campus, The Pennsylvania State University

Patrick Thompson, mathematics education faculty, Vanderbilt University (Nashville, Tenn.)

Bert Waits, emeritus mathematics faculty, Ohio State University

Rose Mary Zbiek, mathematics education faculty, The Pennsylvania State University

Lee Zia, National Science Digital Library, National Science Foundation (Arlington, Va.)

I also communicated by e-mail with Gene Klotz, mathematics faculty, Swarthmore College, and founder of the Math Forum. This set of leaders[1] is,

---

1. Direct quotes cited in this article are taken from verbatim transcripts of either the interviews that were conducted or subsequent e-mails.

of course, hardly a complete set of those who have led the field over the past two decades, but it is a respectable, and I hope representative, subset having a wide-ranging knowledge about technology and teaching. It includes the authors of software that moved the field through interactive geometry tools, dynamic statistics, and computer algebra tools; curriculum writers who have crafted ways to use those tools in mathematics learning; and a publisher who has broken the barriers to widespread adoption of technology in mathematics. It includes the nation's leaders in the use of graphing calculators as well as those who continue to prod the field to move faster and further. It includes individuals who created groundbreaking technology-intensive curricula and individuals who are leading the nation in implementing university and school visions of the use of technology. The interviewees agreed to talk for fifteen to twenty minutes. Most talked for more than twice that amount of time—an indicator of their interest and the depth and breadth of what they had to share.

Some Constructs for Thinking about Enhancing Mathematics Learning through Technology

Jim Fey observed, "Despite the fact that the world hasn't moved where we were ready to move it twenty years ago, it really has changed. We're dealing with generations of people who have different images of mathematics." Part of the reason for this change is that technology has opened up new ways to think about mathematics teaching and learning. We now have constructs for describing aspects of these images. Consider, for example, the myriad routine procedures that students are required to master in school mathematics, procedures that are now well within the capability of handheld technology. In the past, teachers and students were confined to a sequential approach to learning these procedures, with the mastery of each step in the procedure necessary before proceeding to the next step. Technology allows a different approach, with more complex procedures (or *macroprocedures*) chunked into a series of simpler procedures (*microprocedures*) (Heid 2003). For instance, prior to the widespread availability of computer algebra systems or graphing technology, students needed to master factoring polynomials or the quadratic formula before they could consider solving quadratic equations. With technology, students can solve quadratic equations using technological factoring commands before they learn to factor those polynomials by hand, and they can find approximate solutions to quadratic equations in the absence of the quadratic formula by producing calculator-generated graphs. Teachers and students now have choices regarding which procedures to do first and which to do at all. In this type of technological environment, capabilities are distributed between the user and the tool, with the user in charge of making decisions about when and how the tool should be used.

One needs to be careful not to give the impression that technology itself makes the difference in teaching and learning. It is, of course, not the technology that makes the difference but rather how it is used and by whom. Those who have studied the use of technology in mathematics teaching and learning have noted that technology mediates learning. That is, learning is different in the presence of technology. The representations that students access may conceal or reveal different features of the mathematics, and the procedures students assign to the technology (as opposed to doing them by hand) may affect what students process and learn. Moreover, how a student uses technology is dependent on his or her ever-changing relationship to the technology. When a user first encounters a particular technological tool, his or her uses of the technology may be confined to rote application of the specific keystrokes or procedures that had been introduced. As the student develops facility with, and an understanding of, the capabilities of the technology, the technology becomes an instrument that the student can tailor flexibly to specific needs. This development of a working relationship between the user and the tool is called *instrumental genesis*.

Emerging Technologies

Today's new tools and approaches are harbingers of technology in tomorrow's classrooms. A look at some of these emerging technologies through the eyes of the innovators gives some indication of likely futures. Computational tools have the greatest potential for catalyzing specific changes in mathematics classrooms. Tools for communication and information are suited for general technology use in educational settings.

Computational Tools

The interviews identified an array of computational tools for mathematics classes that are likely to influence the future. These tools include geometric construction tools, data analysis tools, computer algebra tools, cognitive tutors, and microworlds. Hardware innovations that will shape the future include a range of handheld configurations of tools, three-dimensional printers, motion detectors, and lighter and more portable "tablet" computers.

Dynamic tools

The type of computational tool that epitomizes visions of the future is the dynamic tool. The most popular dynamic tools of the past decade are the interactive geometric-construction tools like Geometer's Sketchpad (GSP) and Cabri. More recent entries onto the dynamic-tool scene include dynamic data analysis tools—most notably, Fathom. Dynamic tools are characterized by the principles of dynamic manipulation (dragging) and continuous visualization (nearly instantaneous change in the objects on the screen). Rose Zbiek noted that the development of a dragging capacity was a key to the future: "I anticipate advances in dynamic tools in terms of what representations are available and what can be dragged within each representation." Complementing her emphasis on different representations within a tool is an essential difference among dynamic tools in the types of objects that can be changed. Nick Jackiw explained that difference:

> [I]n Sketchpad what is fundamental is the definition of the relationships between objects, and the objects exist as abstractions glued together by these relationships, [whereas] in Fathom … it's instead the collection of data which has a sort of constitutional identity. So you can change the data within that collection and that's … akin to changing the coordinates of something in Sketchpad....

Dynamic tools are likely to be essential components of the technology toolkit in the school mathematics instruction of the future. As dynamic tools take their place in more and more mathematics classrooms, there will be an increasing push for those tools to be combined with other utilities. For example, Jon Choate advocates blending an interactive geometric construction tool with a computer algebra system, and Ken Koedinger sees an intelligent tutor in which an interactive geometric construction tool is embedded. Rose Zbiek identified the importance of this combination of tools:

> It seems to me that the development of hybrid technologies that capitalize on dynamic features and incorporate a blend of current and emerging communication and mathematics utilities is the essence of the future. These tools will combine several of the types of technologies and do so in ways that continue to blur the already deteriorating lines between different areas of mathematics and statistics.

Intelligent tutors and computer algebra systems (CAS)

The computing tools that seem to be most underused in schools today are ones that might be characterized as artificial intelligence tools and ones that perform symbolic manipulation. These tools include computer algebra systems—tools that perform symbolic manipulation as well as generate graphs and perform numerical calculations—and intelligent tutors like the Algebra

Tutor, the Geometry Tutor, and those embedded in the Carnegie Learning curriculum. There is a growing interest among school mathematics leaders in finding optimal ways to incorporate CAS in high school mathematics, exemplified by the annual national conference initiated by a group of Chicago-area teachers collaborating with university faculty (www4.glenbrook.k12.il.us /USACAS/2004.html).

The leaders with whom I spoke had a range of suggestions about what was needed in order for CAS to become a more integral part of the future. Many hoped for a future in which CAS was universally available, pointing out that some CAS capability is already available on the Web. The innovators identified several factors necessary for the promise of CAS to be realized. First, there is a need for curricula that incorporate CAS, not as a dispensable option, but as an integral tool in the development of the mathematics. A start in that direction is Module 9 of the CAS-Intensive Mathematics project (www.ed.psu.edu/casim), a module that uses CAS to focus students on reasoning with and about symbolic representations. (As Rose Zbiek points out, there is a concomitant need for technology-intensive curricula that make use of the range of other emerging technologies.)

Improvements in the CAS tools themselves would make them more viable in the classroom. Jon Choate suggested the development of a CAS that would allow individual teachers to build toolkits tailored to what they were teaching. He described a vector calculator that he has built within Mathematica, with primitives that students can use (e.g., find the length of a vector, find the angle between two vectors, find the cross product, find the dot product). Tools like this would allow the teacher to focus students on particular areas of mathematics at the teacher's discretion. Jon claims that this sort of access to CAS has changed the "playability" of calculus (the extent to which students can get their hands on and investigate calculus concepts) in a somewhat similar fashion to the way that interactive geometry construction tools have changed the playability of geometry. He describes the experience he and his students had with an open-ended but mathematically intense project centering on analyzing carnival rides:

> We got into analyzing carnival rides. We wanted to talk about carnival rides so we got on the web and we went to the company that makes the Mad Hatter ride in Disney World. And there were circles around circles around circles around circles around.... We wondered, "What's going on here?" Well, you come up with functions that describe the situation, but that doesn't tell you why people scream. So that made us think, when are you turning the fastest, when are you getting snapped? With these sorts of questions, you could go to technology and use curvature to analyze what was going on.

This tailoring of the CAS to particular purposes has been a long-term project of Doug Child. He points to his Math T/L (a front end for Maple) and to the home screen on the TI-92 calculator as examples. Both furnish the user with a list of options for viable symbolic manipulations from which the student can choose. This approach focuses the student on a limited range of options. He explains his rationale for tools like this: "So the Math TL and the home screen on the TI calculators can be used to basically focus learning energy. When a teacher wants to teach something, they can think about what the most important aspects of it are and focus on those. The mathematics they want to teach may involve a lot of calculations, but they can factor that out if they want." The calculator can perform the microprocedures (e.g., taking limits, computing derivatives, factoring algebraic expressions) while the student focuses on the higher-level processes, the macroprocedures. Doug explains: "I kept seeing students get focused on little details when they were trying to learn something which I thought was a lot more important and they never even thought about." These tailored and "tailorable" computer algebra systems would allow teachers and students to target instruction more precisely on the specific learning goal.

One of the current features of many CAS is that they do not allow the user to determine the form of the output—a desirable feature, since different symbolic forms reveal different information. The tailorable CAS of the future would also allow users to specify the type of equivalent form in which they would like to see their output. Since tools like this would allow teachers to offer students automated access to any given subset of symbolic manipulations and tools of the future are likely to include access to a range of equivalent but different forms, their use would allow teachers to experiment with different configurations and to learn more about the development of understanding of symbolic representations. Experimentation with these different forms might address the concern expressed by Bert Waits that we do not yet know enough about what paper-and-pencil skills are needed before students can learn mathematics using a CAS. Rose Zbiek deemed the extension of this issue of paper-and-pencil skill to technology use with and beyond CAS as "the number one research issue for the twenty-first century."

The availability of intelligent tutors and CAS tutors inevitably raises the issue of how much guidance to offer students through these tools. Doug Child describes his Symbolic Math Guide (SMG is embedded in TI calculators) as an expert doing symbolic calculations and allowing students the choice of what calculations to do. Ken Koedinger's group works on determining production rules that emulate expert problem solving. With either approach, the question remains concerning the optimal tutoring configuration for student learning.

Microworlds

Ever since Papert's (1993) classic *Mindstorms,* microworlds like Logo have been a staple in the innovative use of computing technology. The extent to which microworlds can become an integral part of mathematics instruction, according to Pat Thompson, depends on the readiness of teachers to adopt a much different view of teaching and learning. He points out:

> Microworlds don't seem to have caught on much. And I think one reason is because the use of microworlds demands a very different conception of mathematical inquiry than most teachers have, because microworlds are typically designed to be experimented with, much like you experiment with some physical system. You poke it here and you poke it there and you try to make it do things and you try to predict how it's going to behave when you try and make it do things. And according to what happens, you increase your understanding of how it works. And that notion of inquiry doesn't fit very well with conceptions of mathematics teaching that most teachers have. Many teachers didn't know what to do with microworlds. They didn't know how to pull mathematics out of them.

This awareness about views of teaching, learning, and mathematics that are needed in order to capitalize on technology is a recurring theme.

Hardware

As the past few decades have shown, the vehicles for implementing software (that is, the hardware) will also be influential. Issues of portability, accessibility, interconnectibility, and interface will be central.

Handheld technology. Chief among hardware products that have made a difference in mathematics teaching is the graphing calculator. John Kenelly hypothesizes: "In terms of fundamentally changing the way people teach, the handheld graphing calculator has been the most influential instrument of all time." It was both what calculators did (expanded visualization capability) and how they did it (in portable, affordable form) that accounts for their unprecedented success. Visions for the future of technology in mathematics education are seeded with ideas about new features being built into handheld devices. These features are likely to have their greatest effect on the dynamics of teaching. Bert Waits believes that the USB ports on TI-84s and wireless computing will have a great impact on the nature of teaching: "The idea of the teacher being able to poll what the students are doing instantly, send stuff back, and perhaps do it wirelessly, that is all very powerful." Others see great potential impact from the data-collection and data-sharing capabilities of handheld devices. Helen Doerr points out that, as mobile computational tools, handheld devices can allow the collection of data that might not otherwise have been collected. Ricardo Nemirovsky sees major impact in the near future from handheld motion detectors with real-time graphical displays. Not all projections are without concern—Lee Zia's optimism about the future potential of handheld and mobile computing is tempered by a recognition of physical limitations. In particular, he points out that "the screen real estate on handheld devices is relatively small."

Alternative hardware configurations for current computers. One can imagine future hardware configurations that address the problems teachers now recognize with current technologies—problems involving accessibility, visibility, portability, and interface. Some of these problems are likely to be addressed by extensions of existing tools. Bert Waits predicts handheld wireless personal computers with bigger and better screens, more memory, and cross-curricular sets of software. Imagine a paper-thin, $8\frac{1}{2}$-by-11-inch, wireless computer equipped with pen input capability, graphic user interfaces, touch screen capability, and powerful computing tools. This would allow students to carry their computers and to use them for a range of classes, not limited to mathematics. Textbooks might exist only as part of each student's personalized set of computer software.

Alternative hardware. Among the likely candidates for new technology in the mathematics classroom are those related to the visualization and creation of three-dimensional objects. One such object, described by Lee Zia, is the three-dimensional printer. Based on devices currently in engineering contexts, the three-dimensional printer would allow the replication of a three-dimensional object through the use of a computer code and appropriate hardware. Lee informally describes the phenomenon, inherent in one of the NSF-funded Digital Library projects, as follows:

> The basic idea is that by creating some kind of high-resolution computer-aided-design (CAD) file, or alternatively, using a high-resolution X-ray computed tomography scanning process, you can produce a huge digital representation that contains all the information about a three-dimensional object: its internal geometry, surface texture, and so forth. Then through basically an inkjet kind of process, layers of material are laid down in successive passes. To create negative spaces, a special chemically treated mixture can be deposited in appropriate places or perhaps some sort of water soluble material, and this part of the object is subsequently washed away. In the end what's left is an actual 3-D object, a complete representation of the original artifact. The printer head just moves over and over, and the digital instructions determine what is deposited, where it is deposited, and how much. In this particular digital library project the artifacts being replicated are one-of-a-kind cast iron models created in the 19th century to aid the teaching and learning of kinematics—which of course provides a context for lots of interesting mathematics!

As these 3-D printers come down in price, students can explore three-dimensional objects in ways similar to how today's students use graphing calculators to explore families of functions. They can alter constraints and see the effects on the resulting objects.

A second category of three-dimension-related hardware that is likely to have an impact on school mathematics is that of 3-D motion detectors. One such device, the Motion Visualizer 3D, is described on the manufacturer's Web site (www.albertiswindow.com/) as a combined hardware-software system that tracks, records, and analyzes an object's motion in three dimensions in real time, based on live video display. Incorporating real-time data capture and immediate graphical display, this tool allows students to experiment with changing parameters in 3-space in much the same way that they now conduct two-dimensional parameter explorations. So, instead of exploring the shape of a two-dimensional graph of a function of the form $f(x) = ax^2 + bx + c$ by examining the effects of changing the values of a or b or c, students can experiment by moving their hands along a surface in order to match a given function rule, thus developing a kinematic sense of the variables (since the student controls things like phase, frequency, and amplitude). The Web site for this product shows students engaging in representing the flight of a free throw and a three-point shot in basketball. Using the 3-D motion detector, they created a graph that displayed the flight of each ball from the initial shot to the point where the ball fell through the basket.

Rose Zbiek pointed out the overarching idea that joins 3-D tools with their 2-D dynamic counterparts when she said: "Three-dimensional and dynamic tools (among other software components) will come together as a larger genre of technology that allows students to test their mathematical ideas by means other than entering syntactically correct literal codes."

Tools for Communication and Information

Much of the focus of current technological progress in education is in the arena of communication and information tools rather than in tools specific to mathematics instruction. These tools are likely to have profound effects on the teaching and learning of mathematics in the near future. This section will discuss the potential in mathematics classrooms for Web-based instruction, for the building of interactive communities, and for research tools.

Web-based instruction

To some people, Web-based instruction seems to be the wave of the future. To many it would seem to offer an economical and viable way to provide instruction to those who might not otherwise be reached. According to the Pew Internet & American Life Project (Madden and Rainie 2003, p. 47), Americans are overwhelmingly drawn to the Internet for educational purposes:

> Over 63% of adult Americans use the Internet.... An average day in 9/02 saw 12 million users taking to the Web in search of information relevant to their educational pursuits.... [S]tudents now rely heavily on the Internet to help them do their schoolwork, some teachers use the Web to facilitate their instruction, and others use the Web in conjunction with continuing education courses or job-related training. Computers with Internet connections are now commonplace in many American class-rooms and some teachers now require some form of online participation from their students.

Not only is the Internet a part of our everyday lives, but it is destined to con-tinue to be so. According to the Pew Internet Project (p. 78):

> The Internet has been irrevocably woven into everyday life for many Americans. While there was once a time when the Internet was interesting because it was dazzling, it is now a normalized part of daily life for about two-thirds of the U.S. population. For some, it has become an integral part of work or school. For others it is a primary means to stay in touch with fam-ily and friends. All the trends set out here seem destined to continue, if not evolve, as the technology gets better, the applications become simpler, the appliances that use the Internet become omnipresent, and the technology fades into the background of people's lives—as powerful, ubiquitous, com-monplace, and "invisible" as electricity.

It is no surprise, then, to see universities turning in droves to exploring the educational and economic feasibility of Web-based instruction. That univer-sities are headed toward delivering complete undergraduate programs on the Web is inevitable, according to Pete Rubba. Because of the lack of availabil-ity of an intuitively easy and stable math editor, however, mathematics-based courses are among the last to be converted. The math editor must be easy to use, since many mathematics students have difficulty with the production of symbolic notation. Rose Zbiek points out an important related learning issue: "Student production of symbols does not immediately follow the con-ventions of formal mathematics. There is a nontrivial task that lies between here and there: creating a math editor that is conducive to student expression in open-ended exploration and problem solving." Other questions that accompany the conversion of mathematics instruction to online include the following: Is instruction online "as good" as it is face-to-face? Will students be able to afford the necessary software and hardware to pursue online mathematics courses? Will online courses adequately address the problems of teaching mathematics in home-school settings or in very small school districts? Will Web-based courses lead to reliance on online quizzes and low-level testing? In spite of these and other issues, Pete Rubba is optimistic about the future of online courses, remarking: "I think we have got to ask ten years from now how much face-to-face instruction will be done at the high school level."

Interactive community building

The use of technology can contribute to interactive community building. For example, with tools like LessonLab, teachers from different districts could collaborate in planning and critiquing mathematics lessons. They could watch the lessons of one another, annotate the videos to highlight areas of potential mutual interest, and meet using videoconferencing equipment. With the capability of completing all this over the Internet, such cross-district collaborations can spread to different states and even to different countries.

The Effects of Emerging Technologies on the Content of School Mathematics

Data analysis and mathematical modeling, now minimal in the school mathematics curriculum, are likely to play a much larger and intertwined role in the school mathematics of the future, partly because of increasingly easy access to increasingly powerful data analysis and function-modeling tools. Moreover, the use of technology in the learning of these areas of mathematics is likely to meet with less resistance than, for example, the incorporation of CAS in the learning of algebra. Jere Confrey offers a possible explanation: "In many ways, that is exactly why the use of software in statistics meets much less resistance—simply because you are dealing with issues of relative precision and accuracy, and people don't think the only story in town is proof."

Many statisticians and mathematicians have carefully distinguished between statistics as a field and mathematics as a field. Bill Finzer explains the role of the technology he has developed in making connections between these fields:

> One way of viewing our work with Fathom is that we are attempting to provide accessible tools that integrate data analysis and mathematics. What does it take to make these tools accessible? We believe that dynamic manipulation (dynamic statistics) is part of it; that when you can reach in and smoothly change the display from one state to another, you have a chance of

> understanding how the representation works and of forming a bond between the data and whatever mathematical model you are starting to make.

He also points to the mathematical work entailed in data analysis of seeing patterns and relationships in data and of building models to describe those relationships. With tools like Fathom, students can download and display data, calculate best-fitting curves, and graph those curves on top of the data. Like other tools (e.g., GSP sketches in CAS-Intensive Mathematics), Fathom supplies dynamically draggable sliders that can serve as parameters for the models being tested.

Jere Confrey points out that the tools are now available for data analysis but that the potential users may not yet be ready: "In some sense, the future is now. Because the tools and data resources are there. The problem is that we're not." She points to the need to develop a kind of reasoning not usually underpinning mathematics courses:

> I think the idea of probabilistic reasoning and reasoning that is much more contingent is going to be of major proportions in the future and people are going to have to begin to grapple with complexity in the sense of not just what is, in a descriptive sense. I think many more people are going to have to begin to understand inferential reasoning and what you can say and not say relative to these complex systems. They will need to develop reasoning that says, "This is the best I can—this is the best guess at this time. And that it may be proven wrong in the future, but I need to go with that kind of contingent reasoning right now."

Jere and others point out that one of the reasons students may not become engaged in mathematics is that they do not generate, or "own," many of the problems on which they work. Jere Confrey describes a project she conducted with her graduate students, examining issues centered on high-stakes testing and on learning enough statistics to investigate the relationships in the data on the basis of looking at distributions and comparisons among groups. With the increasing power and convenience of available data analysis tools, having students generate and pursue their own questions will become viable in a larger number of settings. Fathom, for example, now has a tool for generating surveys and collecting responses.

Technology is likely to change not only the content of school mathematics but also the processes of school mathematics and the nature of mathematical understandings. Students in technologically rich classrooms are likely to develop multirepresentational views of mathematics. Some technologies will enable them to develop almost a kinematic understanding of functional relationships. Ricardo Nemirovsky refers to this kinematic understanding as the development of bodily intuition:

> There is still a bias toward thinking that instruction that is based in formal statements or things that are not sensorial is more elevated or more mathematical or more profound.… There is a huge overlap between what is activated in a brain by thinking about an activity and what is activated when you actually perform that activity. And so I think that for example imagining that a cube rotates in space is deeply rooted in the physical act of rotating cubes with your hand.

Ricardo envisions a future in which there would be a plethora of activities in which students develop a sense of the differences in how mathematical relationships feel.

Technology and the Practice of Teaching

Ideas from the interviews about the potential impact of technology on the mathematics classrooms of the future ranged from claims about the types of teaching and learning afforded by particular technologies to descriptions of classrooms of the future enabled by technologically rich settings and generalization about the types of mathematical thinking supported through particular technological approaches.

Changes in Schools and School Days

Experts came down on both sides of the fence with respect to changes in schools and school days. Pete Rubba, as reported earlier, suggested that Web-based instruction would, in the future, replace much of face-to-face instruction. Bob McCollum, however, is adamant in pointing out the necessity of the teacher in front of students:

> No matter what we do, I feel the most important thing is the teacher in front of the students. I try to be careful to warn people not to get lost in that. In fact, someone just shared a Japanese proverb that says something like: "Better one day with the teacher than a thousand days studying." I just really think that that needs to be at the crux of what we do. But, given that, I certainly recognize and believe that the technology can play a really important role in helping kids to understand things better. That is really what we try to do with it.

Similarly, Kyle Peck makes clear his belief that schools will continue to exist but school days and activities will be different from what they are now:

> You know, some people are probably expecting me to say, well, there won't be any school. There will be school; we have a day-care responsibility—and there is a lot to be said for the collaboration that happens face-to-face.

> Schools should be the most exciting place because there are other kids there
> and because you can be wrestling with cool problems and because you can
> be using tools that you can't necessarily have at every home.

Kyle's work in reforming schools includes the Centre Learning Community
Charter School (the charter school he founded), which he describes as "kids
actively engaged in project-based learning, multidisciplinary, theme-based,
problem-based learning in a technology-rich environment." Continuing to
work in this area, Kyle gives the rationale for the project he is currently
developing:

> So what I envision happening is a school day that's kind of like our day in
> that we have big things we're trying to do and our day is punctuated by
> meetings. So we have these things we're working on and we get up and we
> go and do something else and come back and pick up where we left off. I
> think the student's day is going to be that way in terms of his multidiscipli-
> nary projects. They have these big things that they really are into, using cool
> tools, and we bring them lots of great resources. That way they can create
> things that they are proud of, and their day will be punctuated by skill-based
> lessons—where somebody is trying to teach them something for which they
> have the prerequisites and that they don't already know. Right now when we
> [have] group-based education, we group for the whole semester or the whole
> year. And we put people in groups that turn out to be self-fulfilling prophe-
> cies and so on. So I think what we'll do instead is we'll have just enough
> just-in-time instruction based on an actual understanding of the knowledge
> and skills that people have. So there will be a profile on each student main-
> tained by technology that keeps track of what students know, like medical
> records. I sometimes use a metaphor of a dentist office, because when I go
> to the dentist, they don't say, well, Kyle, you're fifty-one years old, and most
> fifty-one-year-olds are having trouble with their molars, so we're going to
> drill your molars. They say, now let's take a look, and they figure out what
> I need and give me what I need. So I think we've got to do the same thing
> with mathematics skills, but I think the big breakthrough is that the students
> will see math and use math in the context of doing other things.

Kyle sums up the success of CLC by describing comments from one special-
needs student:

> One special-needs kid from the charter school said, "At this school they don't
> do everything for me, because at my last school they used to do everything for
> me. But here they don't. They did a little at first but now they make me do
> everything myself. And I like it that way, and they're much cooler things." And
> to me that quote—"They make me do everything myself, and I like it that way,
> and they're much cooler things."—that really kind of captured it all. I think
> that's what we're going to be able to do with education. We're going to have

kids do things, rather than I tell it to you and you tell it back to me and I give you a grade for it. They're going to have to do things that are cooler things with tools that people outside school are using, with tools that they will take home with them so they will get better and better at these tools because they will do things that they care about in addition to things that we care about. So I think that's what it's about. And math becomes maybe even transparent. They don't even realize when they're doing math and when they're not doing math. They're just building something. And so the math just happens.

Changes in the Role of Students in Defining the Tasks

The issue of students doing tasks they define for themselves is a theme that also arose in other conversations. Steve Rasmussen sees this approach as endemic to dynamic tools. He sees interactive geometry tools as blurring the distinction between the learner and the author of the activities, and he points out that every activity created with Sketchpad can be extended by the user. He contrasts this use of tools with that of microworlds, describing dynamic tools as addressing objectives from the learner's point of view and microworlds as addressing objectives from the teacher's point of view. For example, he sees teachers using Java Sketchpad creating activities from a teacher's point of view, whereas an open-ended use of GSP allows students to define their own tasks. Far less interest is generated in students through these Java Sketchpad predefined types of activities than through GSP. He paints a picture of classrooms in which authority is shared between teacher and students and in which students have the tools to express their mathematical ideas visually. In a school community using GSP, Steve points out that teachers can see what their students can do. In such school settings, there is a breakdown of the expert/nonexpert divide, and the curricula blurs the lines between what students are capable of and what teachers ask them to do. Possible impediments to such an approach in the future are limitations on the availability of the technology or on the willingness of the schools to participate.

Changes in Classroom Connectivity

With the growing popularity of wireless computing and the increasing similarities among computers, calculators, and PDAs, the mathematics classroom is becoming multiply connected—students' technological devices and teachers' technological devices will be connected in ways that allow unprecedented communication. Jim Kaput describes the potential for this kind of connectivity:

> The big old computers, including microcomputers, didn't fit into daily practice very well; they didn't fit into the physical classroom; they didn't fit into

the social structure; and they didn't fit particularly well into the pedagogical structure of most classrooms. The handheld fits much more readily into the physical structure and to a significant extent into the pedagogical structure. While the teacher did not have any kind of direct control or knowledge of what's going on on the handhelds when each of the kids had one, with the new network capacity that is now being built into them, such as the TI Navigator, the teacher can now get what's on the student's handheld as well as broadcast stuff to all the handhelds. With this kind of a network, you can first of all do some traditional things better, more efficiently, like hand out homework, hand out quizzes, collect data, and so on, provided you have the right ancillary software to do that.

Jim describes new "activity structures" for connected classrooms. For example, he describes classroom activities in which students each construct a graph with a given set of constraints—for example, the position versus time graph for two cars in a race. When the students download their graphs to the teacher's computer, they find that the graphs have certain commonalities defined by the constraints and certain differences on characteristics that were free to vary. This activity structure can bring home to students the necessary characteristics of mathematical objects and the characteristics that may vary. The aggregation of the graphs makes salient the defining characteristics of the graphs. Students can, in this way, learn to identify essential and nonessential features of graphs.

The Effect of Technology on Mathematics Learning in Informal Settings

Mathematics learning need not be confined to students or to formal schools and classrooms, and similarly for technology in the service of mathematics learning. The development of structures for informal mathematics learning is in its infancy. Two exemplary projects that illustrate the uses of technology in informal settings are Illuminations and Calculus@Museum.

The Illuminations (illuminations.nctm.org/index.html) project has as a major goal the provision of Internet resources that will help improve the teaching and learning of mathematics for all students. It is intended to provide NCTM *Standards*-based resources for classroom use and professional development for teachers. Eric Hart points out that once these resources are freely accessible on the Internet, they are available in libraries, in community centers, for Parent Nights, and for home schooling. He observes that such sites help the general public to understand what is meant by important mathematics taught in engaging, sense-making ways, what is meant by students learning mathematics in deep ways, and what is meant by teachers teaching mathematics in richer ways. He notes that Illuminations gives examples that can meet all these goals simultaneously.

The Calculus@Museum (www.terc.edu/mathofchange/CM/home.html) project in the Science Museum in St. Paul, Minnesota, focuses on three central concepts: (1) rates of change, (2) incremental summation, and (3) the use of parametric functions to provide a simple way to describe a complex motion. A major goal of this project is to use kinesthetic experiences of physical actions to make calculus ideas accessible and informative to a general audience. Based on TERC's "Line Becomes Motion" (LBM) technology, the project gives visitors "direct experience of these motions through bodily engagement, fully involving the hand as well as the eye in the learning process." Ricardo Nemirovsky describes the exhibit: "You can either draw a graph and drive the physical event or you can move by hand and generate a graph."

Although in the past mathematics has not commonly been an integral part of museums, projects like the two just described that capitalize on emerging technologies furnish exemplars of how technology may affect informal mathematics exhibits over the next decade.

Technology to Help Teachers Do Their Jobs

Technological approaches of the near future will provide welcome assistance to teachers in their day-to-day work. Among the most promising technologies are those that will enable teachers, students, and parents to access the increasing number of technological mathematics-specific tools. Gene Klotz describes his current project, Math Tools:

> A critical challenge is to actually get needed technology materials into the hands of students and teachers. Toward this end, the Math Forum has been experimenting with an online learning community built around a digital library of mathematics software. Our new project, Math Tools, http://math-forum.org/mathtools/, collects math software and organizes it by topic within courses, pre-K–calculus. We also offer user reviews, discussions about the tools, help in using the software, user experiences, problems that use the tools, classroom activities.

Venues like Math Tools afford teachers, mathematics teacher educators, and prospective teachers the opportunity not only to access tools but also to engage in online discussion with other users as well as with the tool developers.

Technology in the mathematics classrooms of the future will not necessarily be entirely mathematics-specific. Some technology will simply be useful to teachers in carrying out the day-to-day responsibilities of their jobs. Bob McCollum and Dan Chazan commented on some of those possibilities:

Logistics of Running a Classroom

- Teachers could take attendance and check in homework on the spreadsheets on their Palm Pilots, then later sit down at their computers and load these records into their electronic grade books.
- Students could beam in their homework.
- Palm Pilots could have probe capabilities for collecting data, and so students could be ready on a moment's notice to collect data for a class experiment.

Providing individual assistance to students

- The Internet could be used as a tutorial medium for students.
- Students could take assessments on the Web, and remediation could be suggested on the basis of their specific answers.

Textbooks and teacher's editions

- Textbooks, although they will still be around, will become less and less important because of the vast array of tools and information that will be available on computers and on Palm Pilots.
- Teacher's editions will be electronic and contain a vast array of resources.
- Teacher's editions could be dynamic, with teachers adding their own activities and comments.

Accountability

- Districts could collect districtwide assessment information online.

Collaboration

- Teachers could share lesson plans and comment on one anothers' plans through shared electronic resources.
- Teachers across districts could participate in teleconferences dealing with common courses and problems.

Regarding this last suggestion, Dan Chazan comments:

> And if you are working electronically, the cost for extra pages [in a teacher's edition] is very small. So if you want to include an essay on the mathematical background behind this task, and you know not everybody is going to need it, in an electronic environment you would be more likely to include it.

> You could have links to the same document from a number of different
> activities in a way that right now in the teacher's guide you can't really do.

Technology also affords new opportunities for professional development.
To Dan, one such role is in supporting communities of teachers: "I can imagine down the line that there will be different tools for supporting communities of teachers to do joint planning of different kinds." To Bob McCollum,
professional development over the Internet helps communication among
teachers in different schools, including those in remote locations:

> I just see the Internet, or whatever will come next from the Internet, as obvi-
> ously being a great way for teachers to access professional development.
> You know, some of the chat room sort of things aren't quite dynamic enough
> and aren't quite real-time enough; but I think they are going to get there; and
> I think we will be able to have my department engage in a discussion with
> your department, for example, over a particular topic, over the Internet,
> cheaply and easily, and the benefits from that I think are huge. I think about
> the teleconferencing capability, too, that you can now do that over the Web,
> and that makes it affordable in a sense. And think about it, especially for
> those schools that can't afford to bring in an expert, but they might be able
> to talk to them over the Internet.

Dan Chazan observes: "Increasingly as people start to use the Web for different kinds of professional development with embedded video technology, for
example, LessonLab and the TIMSS Tools, new ways will be created to talk
about teaching and to read about teaching and to have people communicate
ideas about teaching." A venue for implementing this idea is the VideoPaper
(brp.terc.edu/VPB/vpb.html). In a VideoPaper, "classroom or interview-filmed
episodes can be displayed and synchronized with interpretations, transcriptions, closed captions, images of student work, clarifying diagrams, or other
pieces of information that expand the events, portraying their full complexity.
Teachers, researchers, and other communities interested in video-based dis-
semination can use VideoPapers to make their conversations more grounded in
actual events, more insightful, and more resistant to oversimplifications."

Some Parting Thoughts

Lest this article seem unrealistically optimistic about the technological
future of the mathematics classroom, there are, of course, caveats. If technology is to fulfill its promise, the stars must be at least somewhat aligned. If the
Internet is to continue to provide wonderfully rich resources, there must be
financially viable organizations supporting the creation of such resources. If
teachers are to make the best use of available technologies and tools, they

must be able to locate and take advantage of appropriate professional development. If mathematics classrooms are to become places where students can pursue problems of their own invention and interest, the technology must be available and the curriculum must be open enough to allow such exploration. These changes will take place with teachers who themselves have experienced learning mathematics with technology.

Nevertheless, according to this amazing group of national leaders in education, the many possible futures of technology in mathematics education are bright and enticing. An essential ingredient of some visions of the future is that we will be using well and regularly the technologies that now exist (but are not used to any significant extent in today's classrooms). There were some important ideas that came to the fore in the course of my hours of conversations with these educators. I conclude this article with a list of principal predictions associated with the range of possible futures for technology and mathematics instruction:

- Wireless connectivity suggests a future in which students and teachers freely share and build on the ideas of one another.

- Students will experience mathematics kinematically.

- Web-based facilitation of learning will grow and develop.

- New kinds of reasoning (rooted in reasoning in the face of uncertainty and reasoning with contingencies) will be needed as technology-supported data analysis takes its rightful place in the school mathematics curriculum.

- Teachers will use just-in-time technology-based instruction for honing students' missing skills.

- New school configurations will allow for smooth transitions between technology-assisted group work and technology-delivered individualized instruction.

- Teachers and students will regularly use tailored and tailorable computer algebra systems and intelligent tutors.

- Technology at its best will allow students to test their mathematical ideas by means other than entering syntactically correct literal codes.

As Rose Zbiek pointed out in our conversation, the coming decade of technology use in mathematics classrooms will be punctuated by the growing need to understand the intricate relationship between doing anything by hand or by technology and the development of students' understanding at a more global level. It is this shared commitment to intense use of technology not for the sake of using tools but for the sake of developing and enhancing students' mathematical understanding that will serve the future well.

REFERENCES

Heid, M. Kathleen. "Theories for Thinking about the Use of CAS in Teaching and Learning Mathematics." In *Computer Algebra Systems in Secondary School Mathematics Education,* edited by James T. Fey, Al Cuoco, Carolyn Kieran, Lin McMullin, and Rose Mary Zbiek, pp. 33–52. Reston, Va.: National Council of Teachers of Mathematics, 2003.

Madden, Mary, and Lee Rainie. "America's On-Line Pursuits: Who's On Line and What They Do." Washington, D.C.: Pew Internet & American Life Project, December 28, 2003. (www.pewinternet.org/reports)

Papert, Seymour. *Mindstorms: Children, Computers, and Powerful Ideas.* New York: Basic Books, 1993.